W9-AWE-463

NATIONAL GEOGRAPHIC

SCIENCE

SCIENCE INQUIRY
AND WRITING BOOK

SCIENCE

NATIONAL GEOGRAPHIC
School Publishing

PROGRAM AUTHORS

Judith S. Lederman, Ph.D.
Randy Bell, Ph.D.
Malcolm B. Butler, Ph.D.
Kathy Cabe Trundle, Ph.D.
David W. Moore, Ph.D.

Science Inquiry

Life Science

Life Science

Science Inquiry

Earth Science

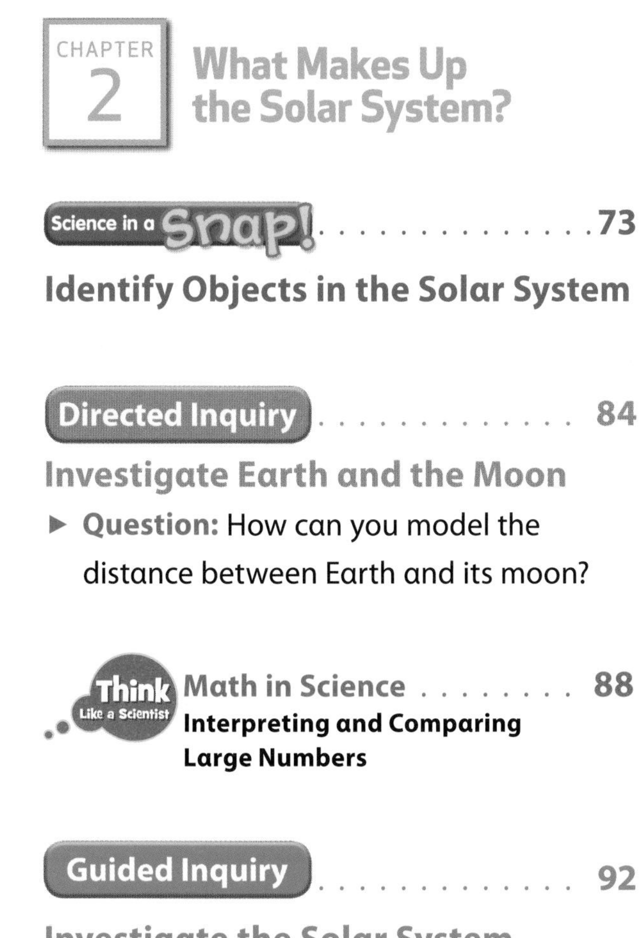

Science Inquiry

Earth Science

How Can We Protect Earth's Resources?

CHAPTER 5

How Are Weather and the Water Cycle Connected?

Physical Science

How Can You Describe Matter, Mixtures, and Solutions?

CHAPTER 2

How Can Matter Change?

Science Inquiry

Physical Science

Physical Science

Science in a Snap!

CHAPTER 1

Science in a Snap! Observe and Classify Animals

Make a list of all the animals you might **observe** in a single day. When your list is complete, think about the body structure of each animal. **Classify** each animal as a vertebrate or an invertebrate. What characteristics do all the vertebrates on your list share?

Science in a Snap! Explain How Organisms Interact

Some organisms have a relationship in which they benefit each other. **Observe** the photographs of the plants and animals on the page. Match the organisms that have a beneficial relationship. How does each organism get what it needs to survive?

Oxpecker needs food, eats insects.

Ant needs thorns to make a home.

Apple flower needs pollination.

Twig of acacia tree needs protection from plant eaters.

Honeybee needs food, collects nectar.

Water buffalo needs insects removed from skin.

CHAPTER 3

Science in a Snap! Make a Model of a Food Web

A pond ecosystem has many kinds of living things. Write **sun** on an index card. Then write the names of these pond organisms on separate index cards: **tadpole, frog, pond snail, diving beetle, hawk, heron, algae, dragonfly, crayfish, minnow (small fish), bass (large fish), moss, grass, goose, raccoon, floating plants.**

Place each organism card so that it is touching the card or cards for its source of energy. You have made a food web. Identify the producers and consumers in your web.

CHAPTER 4

Science in a Snap! Classify Animal Behaviors

Think of a pet or another animal that you have **observed.** Make a list of its behaviors. **Classify** each behavior as either an instinct that the animal inherited from its parents or as a behavior that it learned. **Compare** your list with a classmate who chose a different animal. Did the two animals have any behaviors in common? Which behaviors help the animals survive in their environments?

CHAPTER 5

Science in a Snap! Make a Model Eardrum

Stretch a piece of plastic wrap tightly across the top of a plastic cup. Secure the plastic wrap around the cup with a rubber band. Sprinkle a small amount of salt on the plastic wrap. Hold a bowl close to the plastic wrap and hit the bowl several times with a spoon. **Observe** what happens to the salt. What caused the salt to move? Explain how the plastic wrap acts as a **model** of an eardrum. Why might you use a model to learn more about eardrums?

Explore Activity

Investigate Identification Keys

How can you identify plants and animals using identification keys?

Science Process Vocabulary

observe verb

When you **observe,** you use your senses to gather information about an object or event.

classify verb

When you **classify,** you put things in groups according to their characteristics.

Materials

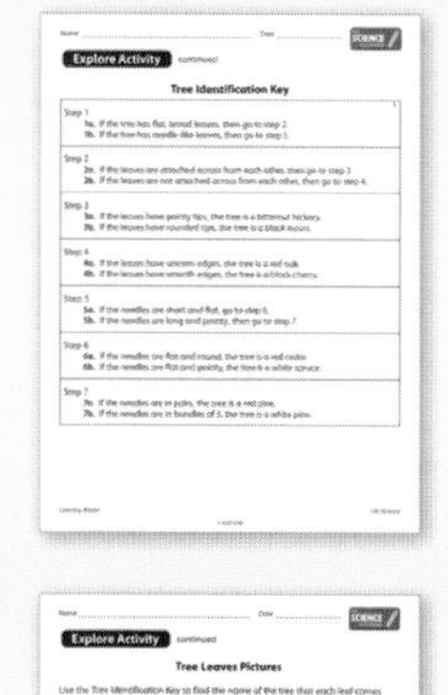

Tree Identification Key

Tree Leaves Pictures

Fish Identification Key

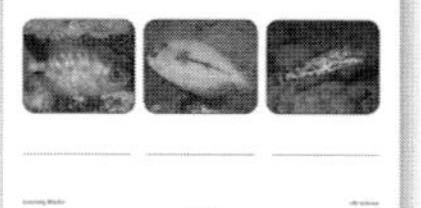

Fish Pictures

What to Do

1. **Observe** the characteristics of the tree and its leaves in the pictures. You will use your observations and the Tree Identification Key to identify the tree.

2. Read step 1 of the Tree Identification Key. **Classify** the tree's leaves as either flat and broad or needle-like. Record your choice in your science notebook. Then follow the directions on the key to continue to the next step.

3. Follow the remaining steps on the Tree Identification Key until you can identify the tree. Record the characteristic that you match to the tree at each step.

What to Do, continued

4 Look at the other leaves on the Tree Leaves Pictures Learning Master. Use the Tree Identification Key to identify each kind of tree by its leaves. Make a new table on a separate sheet of paper for each tree.

5 Observe the fish in the picture below. Follow the steps on the Fish Identification Key to identify the fish. Record your choice at each step.

6 Then use the Fish Identification Key to find the name of each fish on the Fish Pictures Learning Master. Make a new table for each kind of fish.

Record

Write in your science notebook.
Use tables like these.

Tree Identification

Step Followed	Characteristic
1	

Fish Identification

Step Followed	Characteristic
1	

Explain and Conclude

1. What is the name of the tree on page 15? What is the name of the fish on page 16?
2. Explain how you used the identification keys to identify trees and fish.
3. **Share** your results with other groups. Did all groups identify the trees and the fish the same? Explain any differences.

How could you use a key to identify this fish?

Investigate Plant Cells

How do onion cells compare with *Elodea* cells when you observe them under a microscope?

Science Process Vocabulary

observe verb

You can use equipment, such as a microscope, to **observe** objects more closely.

compare verb

When you **compare,** you tell how objects or events are alike and different.

How are these plant cells alike and different?

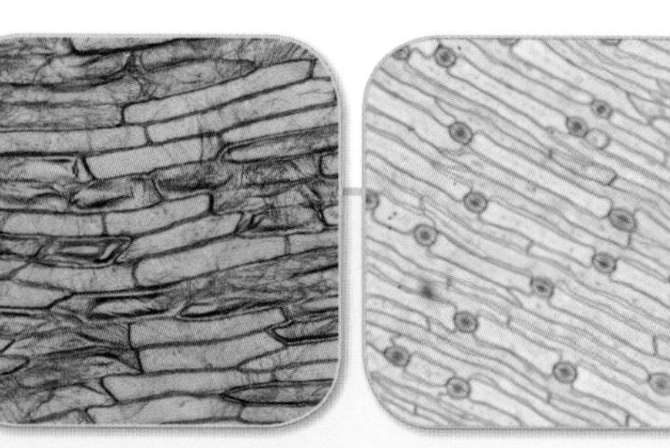

Materials

onion cells slide

microscope

Elodea cells slide

What to Do

1. Place the onion cells slide on the stage of the microscope. Look through the eyepiece and adjust the mirror until you see the most light. Then turn the adjustment knob until you have an onion cell in focus.

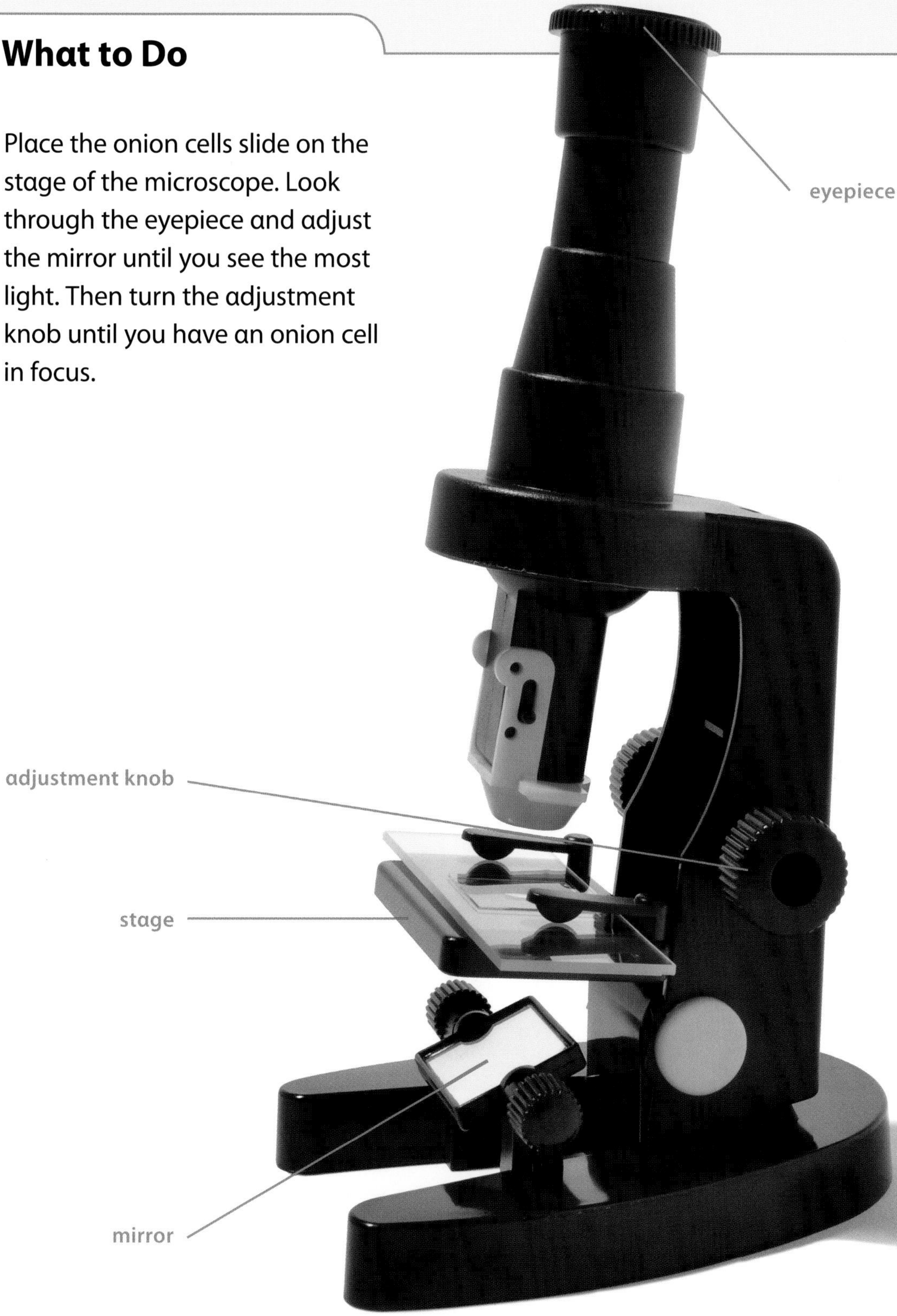

What to Do, continued

2. **Observe** the onion cells through the microscope. Record and draw your observations in your science notebook.

3. Use the picture of the onion cells below to label any cell parts you observed.

Onion cells

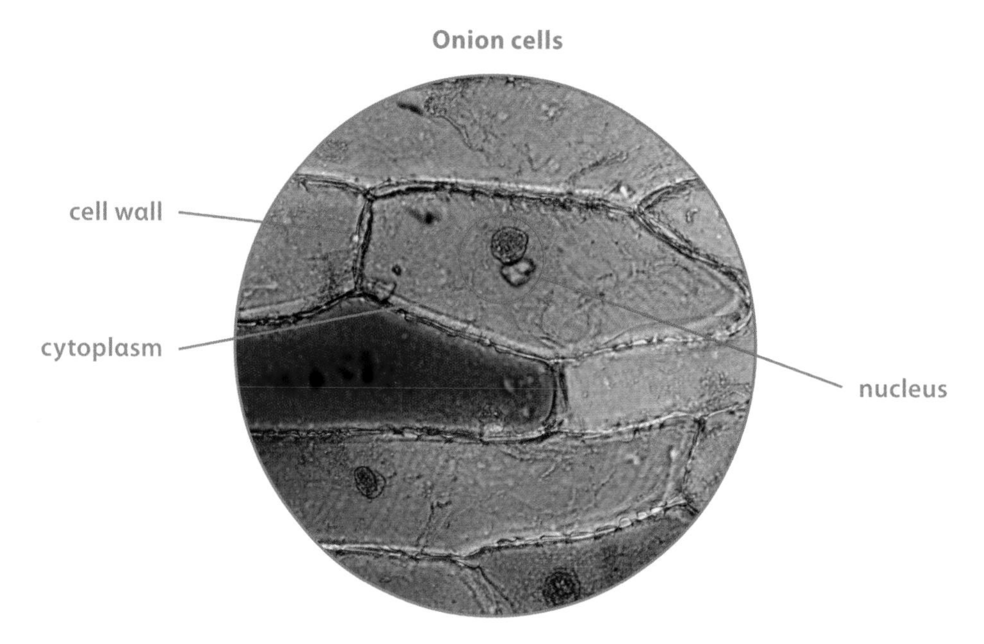

4. Repeat steps 2 and 3 using the *Elodea* cells slide and the picture of the *Elodea* cells below.

***Elodea* cells**

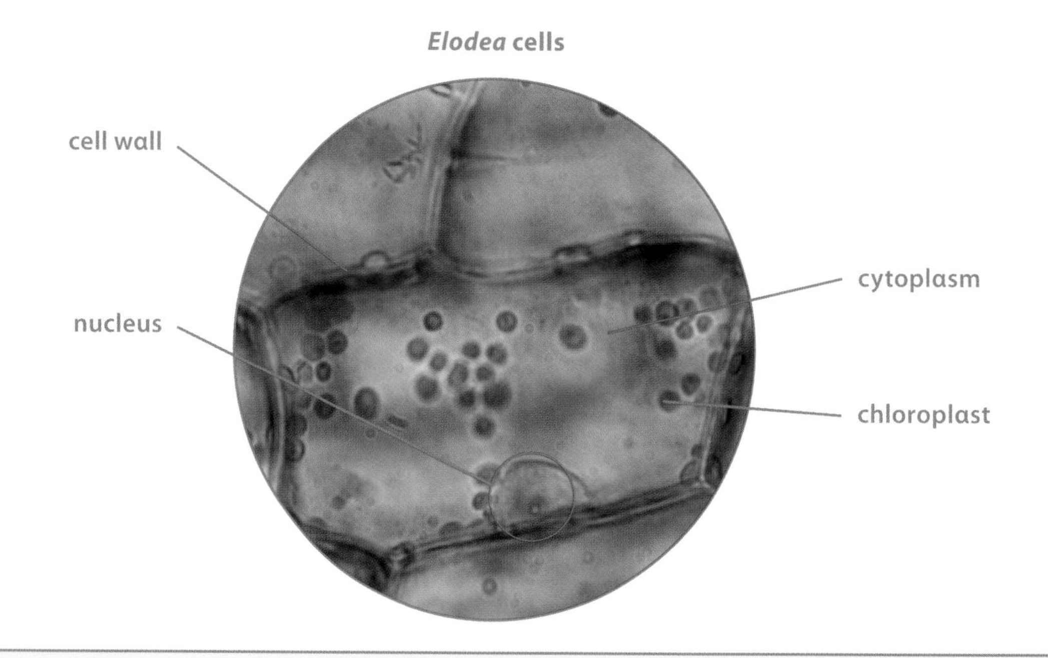

Record

Write and draw in your science notebook.
Use a table like this one.

my SCIENCE notebook

Plant Cells Observed with a Microscope

Cell	Observations
Onion	
Elodea	

Onion Cell

Elodea Cell

Explain and Conclude

1. What structures did you **observe** in the onion cells?
   What structures did you observe in the *Elodea* cells?
2. **Compare** your drawings of the onion cells and the *Elodea* cells.
   How are the cells alike and different?
3. The onion cells come from an onion bulb, which is where food is stored. The *Elodea* cells are from a leaf of an *Elodea* plant. Why do you think the *Elodea* cells have chloroplasts and onion cells do not?

Think of Another Question

What else would you like to find out about *Elodea* and onion cells? How could you find an answer to this question?

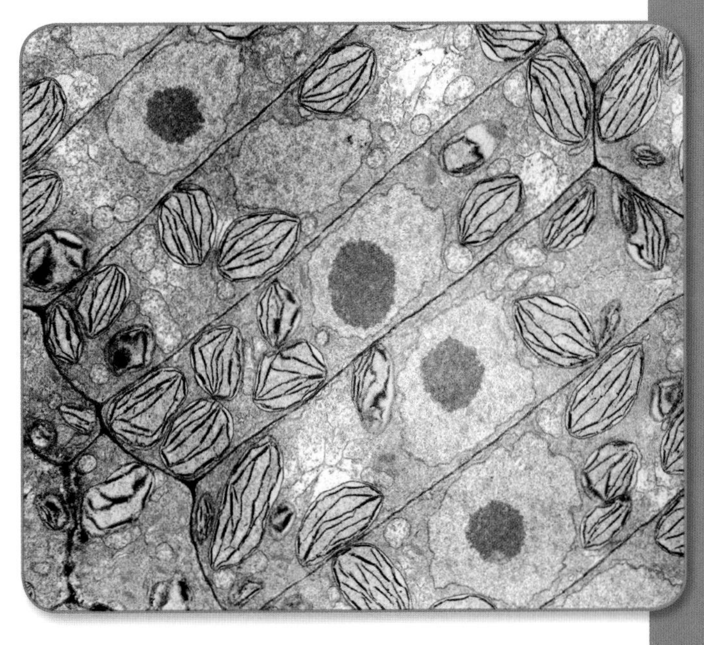

These duckweed cells are magnified 10,900 times.

Directed Inquiry

Investigate Changing Ecosystems

Question **How can you model the change in a pond ecosystem over time?**

Science Process Vocabulary

model noun

A **model** can show how a process works in real life. For example, a model can show how an ecosystem changes over time.

conclude verb

You **conclude** when you use your observations to come up with a decision or answer.

I will use my observations to conclude how ecosystems can change.

Materials

safety goggles

plastic container

measuring cup

potting soil

water

Elodea plants

plastic spoon

rye grass seeds

clover seeds

daisy seeds

humus

What to Do

1. Put on your safety goggles. Evenly cover the bottom of the plastic container with about 3 cm of soil. Then slowly pour 4–5 cups of water on top of the soil. You have made a **model** pond. Allow your model to sit out overnight.

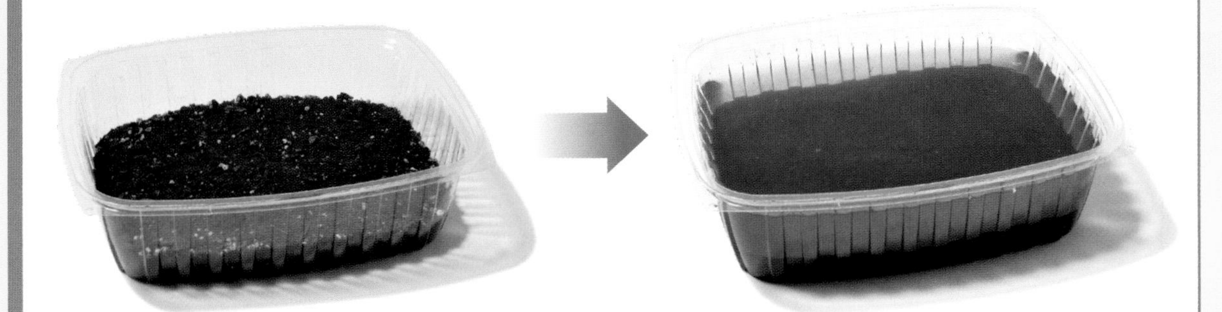

2. The next day, gently place the *Elodea* plants in the water. Then use the spoon to sprinkle 5 seeds each of rye grass, clover, and daisy over the model pond. Put your pond in a sunny place. Record your **observations** in your science notebook.

What to Do, continued

3. Wait 3 days. Then sprinkle a small amount of seeds and 3 spoonfuls of humus on the model pond. The humus models water plants and other organisms that die and decay in the pond. Record your observations.

4. Continue sprinkling seeds and humus and observing every 3 days for about 2 weeks. Record your observations. Be sure to note any changes to the water, soil, and plants.

Record

Write and draw in your science notebook.
Use a table like this one.

my SCIENCE notebook

Pond Ecosystem Model

Date	Observations

Explain and Conclude

1. How is your **model** like a real pond ecosystem over time? How is it different?
2. What changes did you **observe** in the pond ecosystem model over 2 weeks?
3. What can you **conclude** caused the change to the kinds of plants that grew in the pond ecosystem over time? Use evidence from the **investigation** to support your answer.

Think of Another Question

What else would you like to find out about how you can model changes to an ecosystem? How could you find an answer to this new question?

Apostle Islands, Wisconsin

This marsh was once a pond. Conditions changed and marsh plants replaced pond plants. As the ecosystem changes, meadow plants may grow.

Guided Inquiry

Investigate Pollution and Plants

Question **How does pollution in water affect the growth of rye grass?**

Science Process Vocabulary

hypothesis noun

You make a **hypothesis** when you state a possible answer to a question that can be tested by an experiment.

My hypothesis is that pollution such as acid rain will cause damage to plants.

experiment noun

In an **experiment,** you change only one variable, measure or observe another variable, and control other variables so they stay the same.

I will design an experiment to find out if my hypothesis is supported.

Materials

safety goggles

graduated cylinder

water

2 cups

tape

salt water

vinegar

oil

2 cups of rye grass

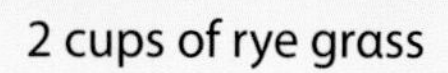

spoon

ruler

Do an Experiment

Write your plan in your science notebook.

Make a Hypothesis

In this investigation, you will observe the growth of 2 rye grass plants in 2 different cups. You will water 1 cup of grass with plain water. You will water the other cup of grass with water that is polluted. You can choose oil, vinegar, or salt as your pollutant. How will the pollutant affect the growth of the rye grass? Write your **hypothesis.**

Identify, Manipulate, and Control Variables

Which variable will you change?
Which variable will you measure or observe?
Which variable will you keep the same?

What to Do

1. Put on your safety goggles. Use the graduated cylinder to pour 75 mL of water into each of the 2 cups. Then label 1 cup **Water.** You will not add a pollutant to the Water cup.

2. Choose a pollutant. Label the cup of water to tell which pollutant you are adding. Record your choice in your science notebook. Then use the graduated cylinder to **measure** 20 mL of the pollutant and add it to the cup.

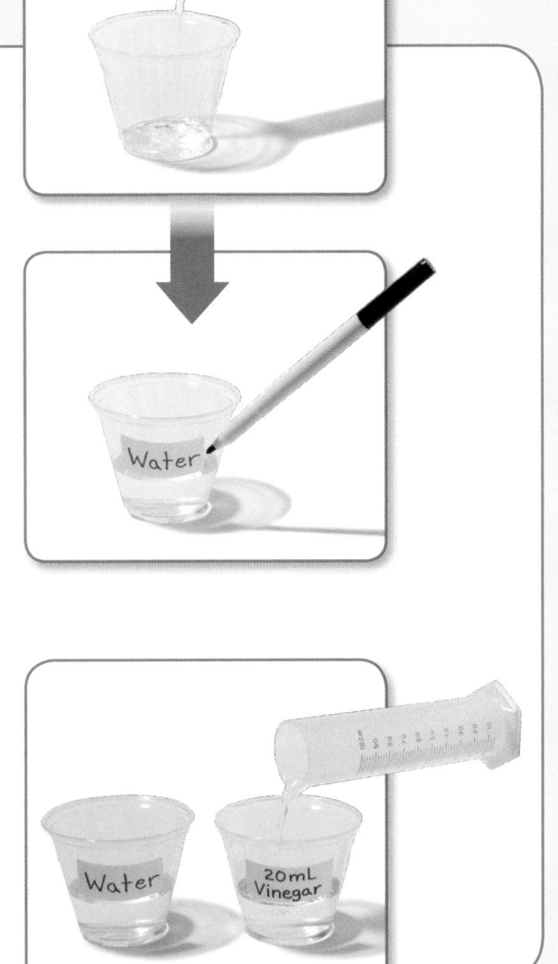

What to Do, continued

3. Label 1 cup of rye grass **Unpolluted Water**.
Label the other cup of rye grass **Polluted Water**.

4. **Observe** the rye grass in each cup. Use the ruler to measure the height of the grass in each cup. Record your measurements and other observations in your science notebook.

5. Add 2 spoonfuls of water to the grass in the Unpolluted Water cup.
Add 2 spoonfuls of the polluted water to the grass in the Polluted Water cup.

6. Repeat step 5 every day for 2 weeks. Observe and measure the rye grass in each cup every other day. Record your observations and measurements.

Record

Write in your science notebook.
Use a table like this one.

my SCIENCE notebook

Growth of Rye Grass

		Unpolluted Water Cup (Water Only)	Polluted Water Cup
Day 1	Height of rye grass (cm)		
	Observations		

Explain and Conclude

1. Did your results support your **hypothesis?** Explain.
2. How did adding the pollution affect the growth of the rye grass?
3. **Compare** your **data** with other groups. Which pollutant had the greatest effect on the rye grass?

Think of Another Question

What else would you like to find out about how pollution in water affects the growth of plants? How could you find an answer to this new question?

Willamette Valley, Oregon

In this area of Oregon, conditions for growing rye grass are very good.

Investigate Producers and Consumers

Where does the energy in a pill bug's food come from?

Science Process Vocabulary

observe verb

When you **observe,** you use your senses to learn about an object or event.

I can observe the lettuce plants to see how they grow.

classify verb

You **classify** when you put things into groups according to their characteristics.

I will classify the lettuce plant as a producer or a consumer.

Materials

safety goggles

spoon

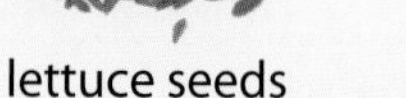
lettuce seeds

cup with soil

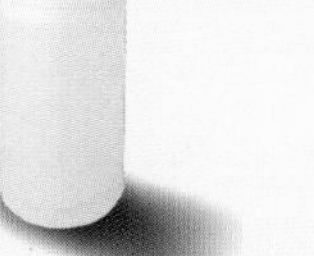
spray bottle

scissors

paper plate

plastic container

paper towels

5 pill bugs

What to Do

1. Put on your safety goggles. Use the spoon to plant the lettuce seeds in the soil. Make sure that soil covers the seeds, and gently press down the soil. Use the spray bottle to moisten the soil. Place the cup with seeds in a sunny place.

2. **Observe** the cup each day. Record your observations in your science notebook. Use the spray bottle to keep the soil moist.

3. When the lettuce plants grow leaves, use the scissors to carefully cut the leaves from the plant. Cut the leaves into small pieces and place the pieces on a paper plate.

What to Do, continued

4. Cover the bottom of the plastic container with paper towels. Spray the paper towels until they are moist. Place the lettuce leaf pieces in a pile at one end of the container.

5. Use the spoon to gently add 5 pill bugs to the container. Observe the behavior of the pill bugs. Record your observations.

6. Continue to observe the pill bugs and leaf pieces every day for 3 more days. Use the spray bottle to keep the paper towels damp. Record your observations.

Record

Write and draw in your science notebook. Use tables like these.

my SCIENCE notebook

Lettuce Plants

Day	Observations
1	

Pill Bugs and Lettuce Leaves

Day	Observations
1	

Explain and Conclude

1. Use your **observations** and what you know about plants to **infer** how the lettuce plant gets energy.
2. Based on your observations, explain where the energy comes from in the food that pill bugs eat. **Classify** the lettuce plants and the pill bugs as producers or consumers.

Think of Another Question

What else would you like to find out about how pill bugs get energy from the sun from their food? How could you find an answer to this new question?

How do these corn plants get energy to live and grow?

Guided Inquiry

Investigate Mold

How can you use observations of mold to infer where it gets its energy?

Science Process Vocabulary

compare verb

When you **compare,** you tell how objects or events are alike and different.

infer verb

When you **infer,** you use what you know and what you observe to draw a conclusion.

Materials

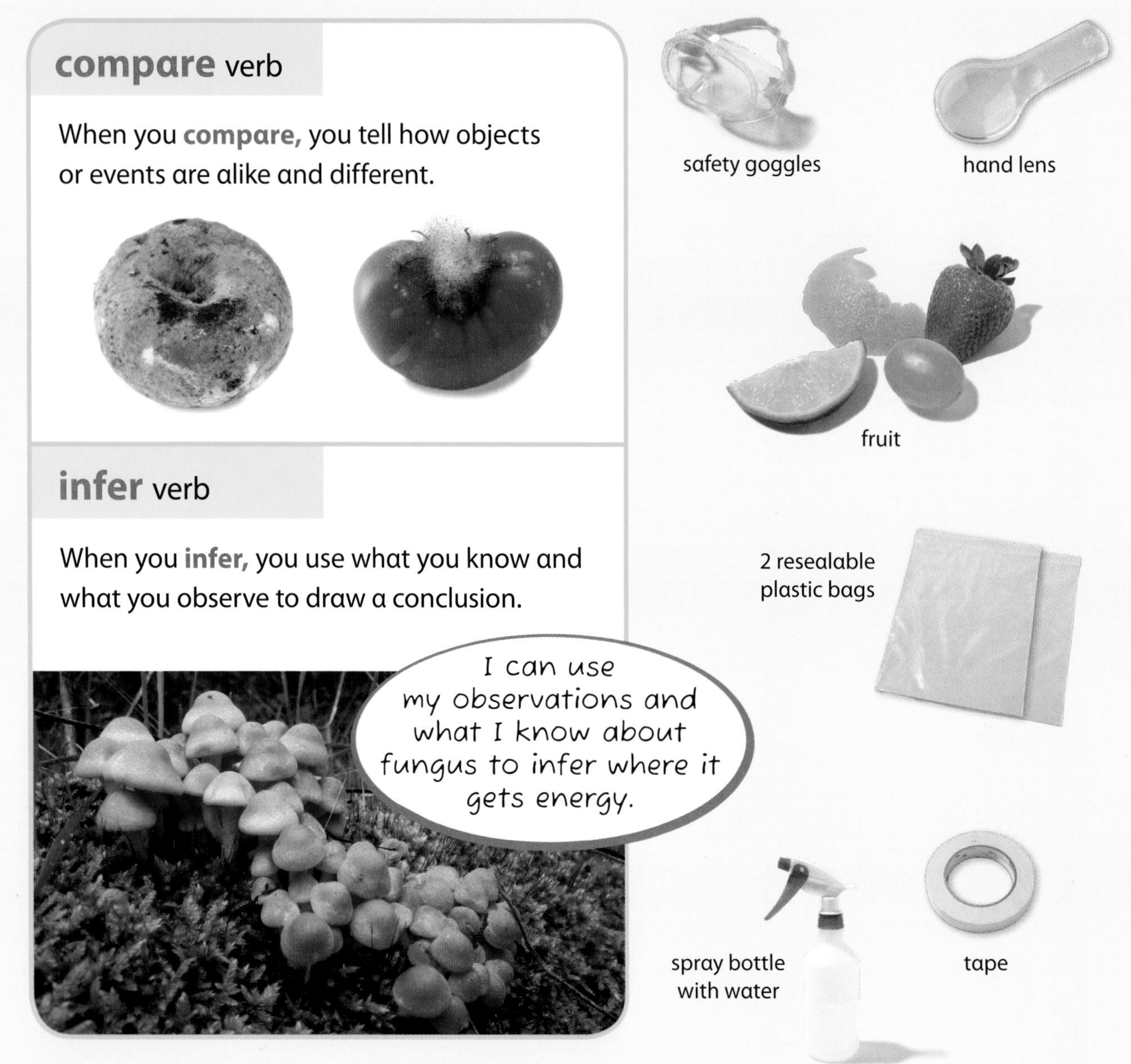

What to Do

1. Put on your safety goggles. Choose which type of fruit you will **investigate.** Record your choice. Use the hand lens to **observe** the surface of the fruit. Record your observations in your science notebook.

2. Carefully place the fruit in a plastic bag. Lightly spray the fruit with water.

3. Seal the bag. Then place the bag with fruit inside another plastic bag. Tape the second bag shut.

Make sure both bags are sealed tightly.

What to Do, continued

4. Place the bags with fruit on a flat surface in the shade. Wait 1 day and observe the fruit inside the bags. Do not open the bags. Record your observations.

5. Observe the bags every day for 1 week. When mold begins to grow on the fruit, be sure to note the color and texture of the mold. Record the amount of mold. Do not open the bags. Record your observations each day.

Record

Write and draw in your science notebook.
Use a table like this one.

my SCIENCE notebook

Observations of ______________________

Day	Observations
Start	
1	

Explain and Conclude

1. Mold is a living thing. Use your results to **infer** where mold gets its energy. Explain.
2. **Compare** your observations with other groups that investigated the same fruit. Compare your results with groups that used different fruits. **Analyze** your data by looking for patterns.
3. Draw a food web that includes your fruit and mold. Label the producers, consumers, and decomposers in your food web.

Think of Another Question

What else would you like to find out about mold and food webs? How could you find an answer to this new question?

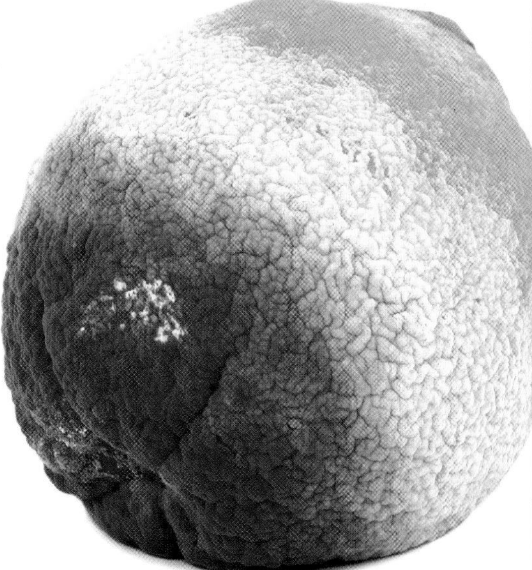

Mold is decomposing this orange.

Math in Science

Recording and Interpreting Data

Scientists collect data as they conduct investigations. They record their observations and data in a variety of ways. Scientists need to keep accurate records during an investigation so that they can draw conclusions from their results.

Scientists often organize their data in charts, tables, or graphs. They may also choose to make diagrams or drawings to record their observations. Scientists decide how to organize their data based on the data and the question they are trying to answer.

Charts and Tables Information in charts and tables is easy to read. Information is organized in columns and rows. Headings for each column and row tell what data is included in each part. Scientists often find it useful to record data in a table so that they can compare results from different trials.

Earthworms

	Mass (g)	Length (cm)
Worm A	4	11
Worm B	3	9

Graphs Scientists often display the results of an investigation in a graph to make it easier to see patterns in the data. Graphs can help you interpret the results of a scientific investigation.

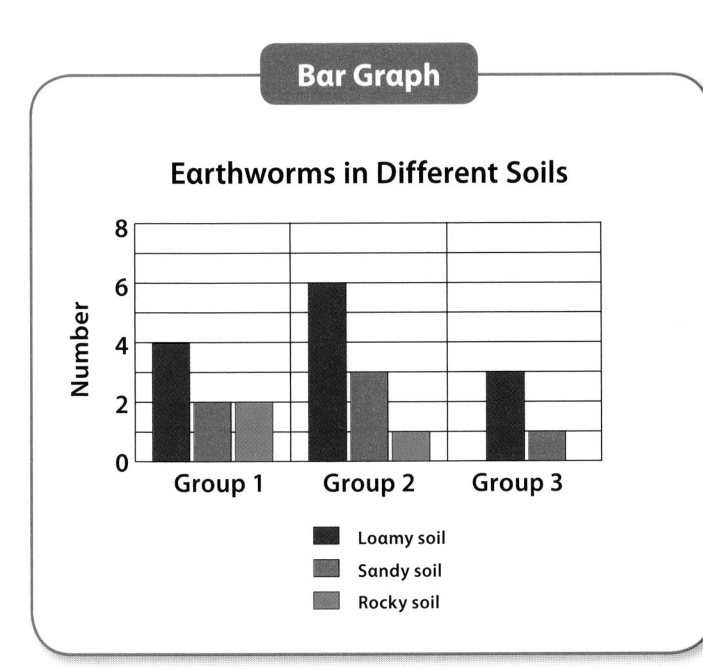

A bar graph can help you compare data that was taken from different groups or at different times.

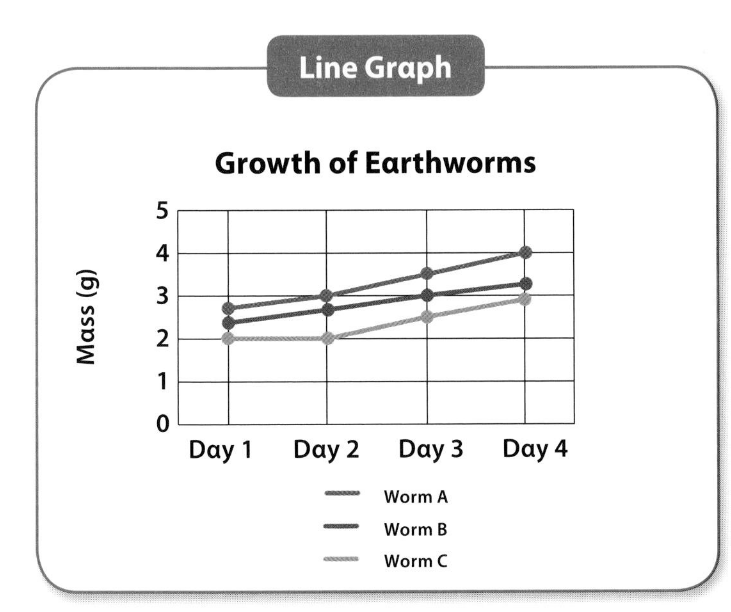

A line graph is helpful in showing change over time.

Drawings and Diagrams Scientists often find it helpful to draw their observations during an investigation. Using drawings and diagrams to record data helps scientists to remember and analyze what they have observed.

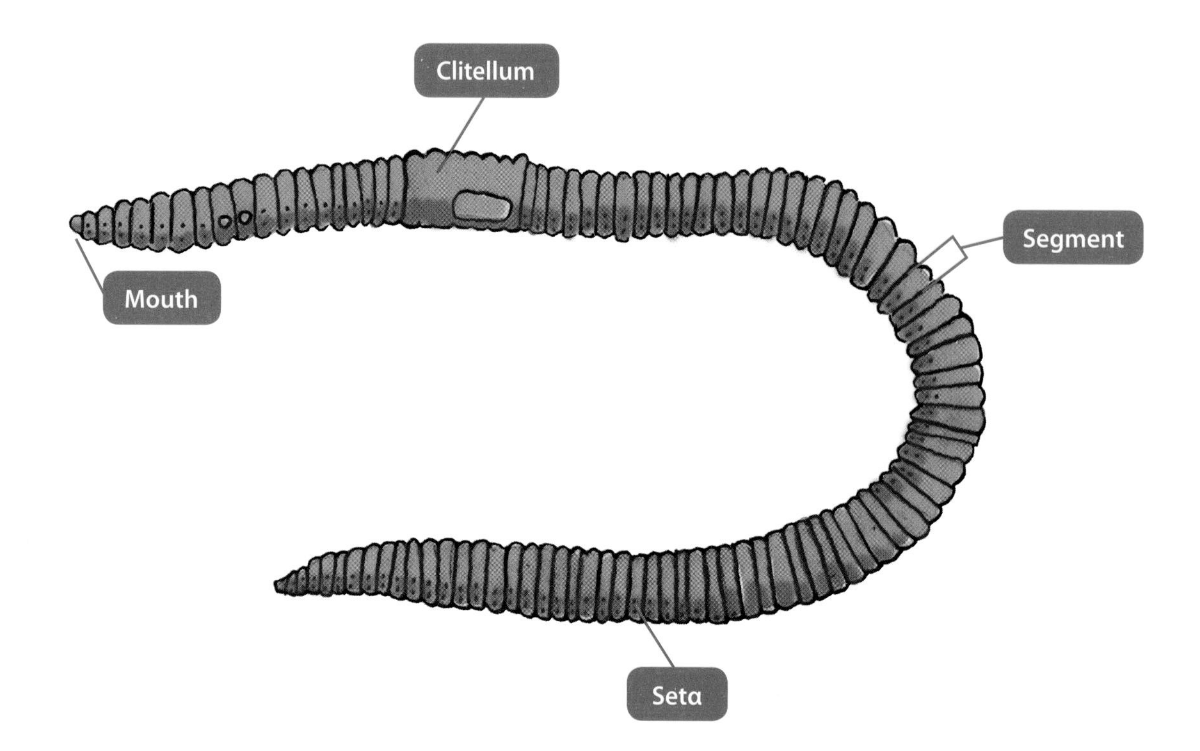

SUMMARIZE

What Did You Find Out?

1. Why do scientists sometimes organize their data in tables or charts?

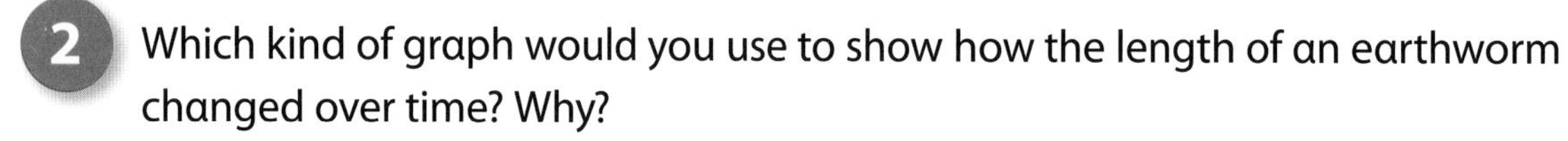

2. Which kind of graph would you use to show how the length of an earthworm changed over time? Why?

3. How can a diagram help a scientist interpret results of an investigation?

Interpret Data in a Line Graph

The line graph below shows data about the earthworm population in a particular place.

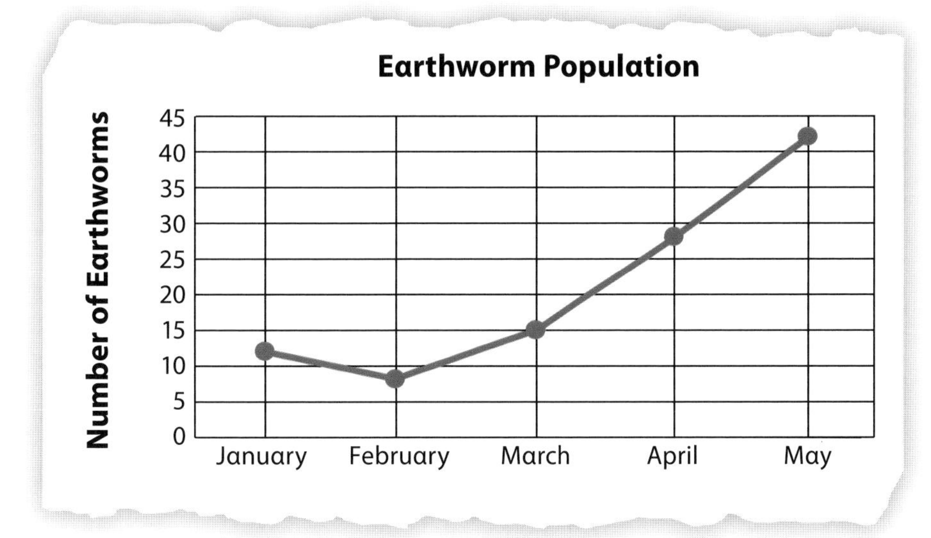

1. Look carefully at the line graph. Read the title and labels.

2. What information is being presented in the graph?

3. What conclusions can you draw about the earthworm population from the data in this graph?

Directed Inquiry

Investigate Bird Feathers

How do different types of feathers help birds survive?

Materials

3 feathers

hand lens

ruler

fan

dropper

cup with water

observe verb

When you **observe,** you use your senses to learn about an object or event.

infer verb

When you use what you know and what you observe to draw a conclusion, you **infer.**

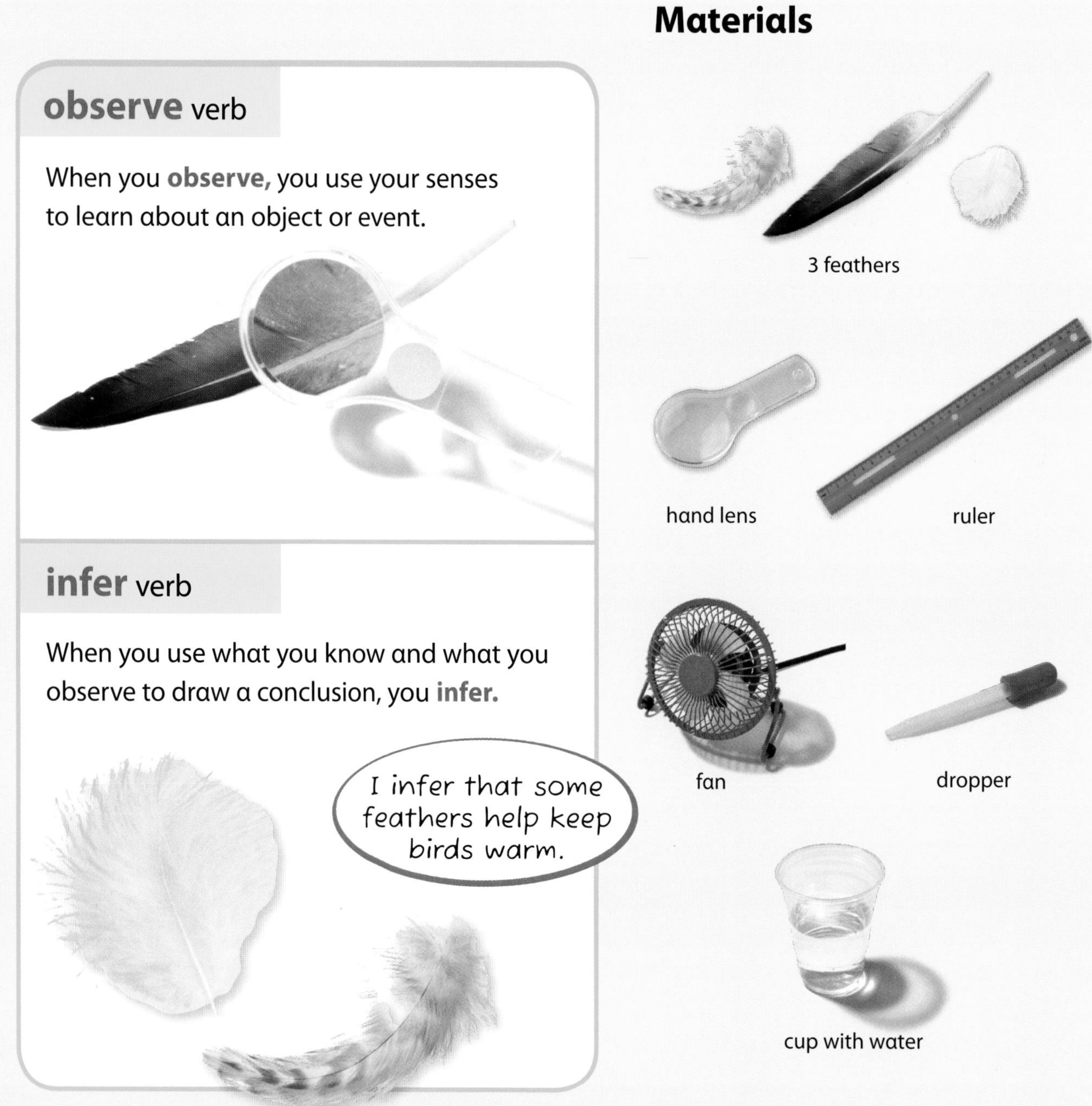

What to Do

1. Choose a feather and **observe** it with a hand lens. Describe the feather, including its shape, color, and texture. Draw your feather and record your observations in your science notebook.

2. **Measure** the length of the feather with a ruler. Record your **data.**

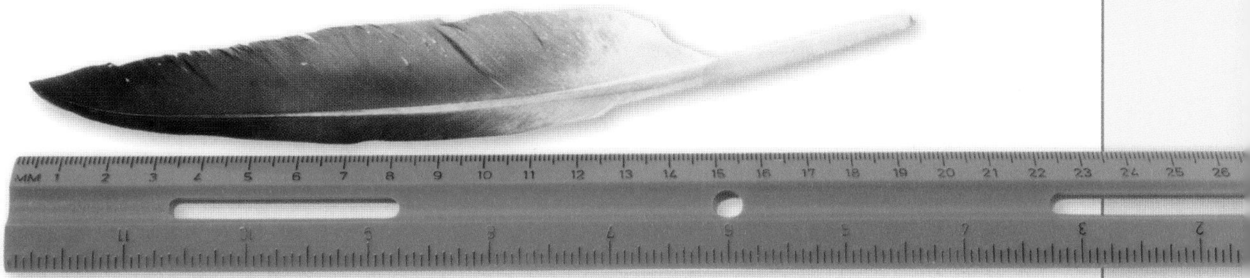

What to Do, continued

3 Hold the feather in front of a fan and observe what happens. Move the feather in different directions. How does the feather react to the wind from the fan? Record your observations.

4 Use the dropper to place 3–4 drops of water onto the feather. Record your observations.

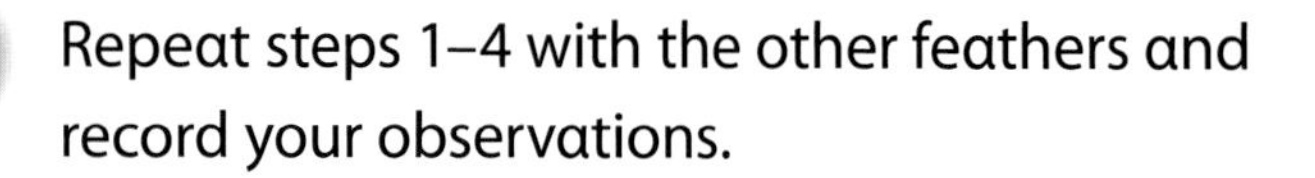

5 Repeat steps 1–4 with the other feathers and record your observations.

Record

Write and draw in your science notebook.
Use a table like this one.

my SCIENCE notebook

Bird Feathers

Draw Your Feather	Shape, Color, and Texture	Length (cm)	When Held in Front of Fan	When Water Dropped on It

Explain and Conclude

1. **Compare** the 3 feathers.
2. Use your observations to **infer** how each feather helps a bird survive in its environment.

Think of Another Question

What else would you like to find out about how different types of feathers help birds survive? How would you find an answer to this new question?

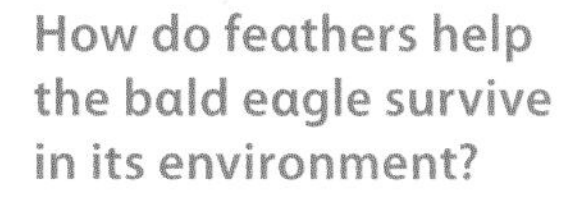

How do feathers help the bald eagle survive in its environment?

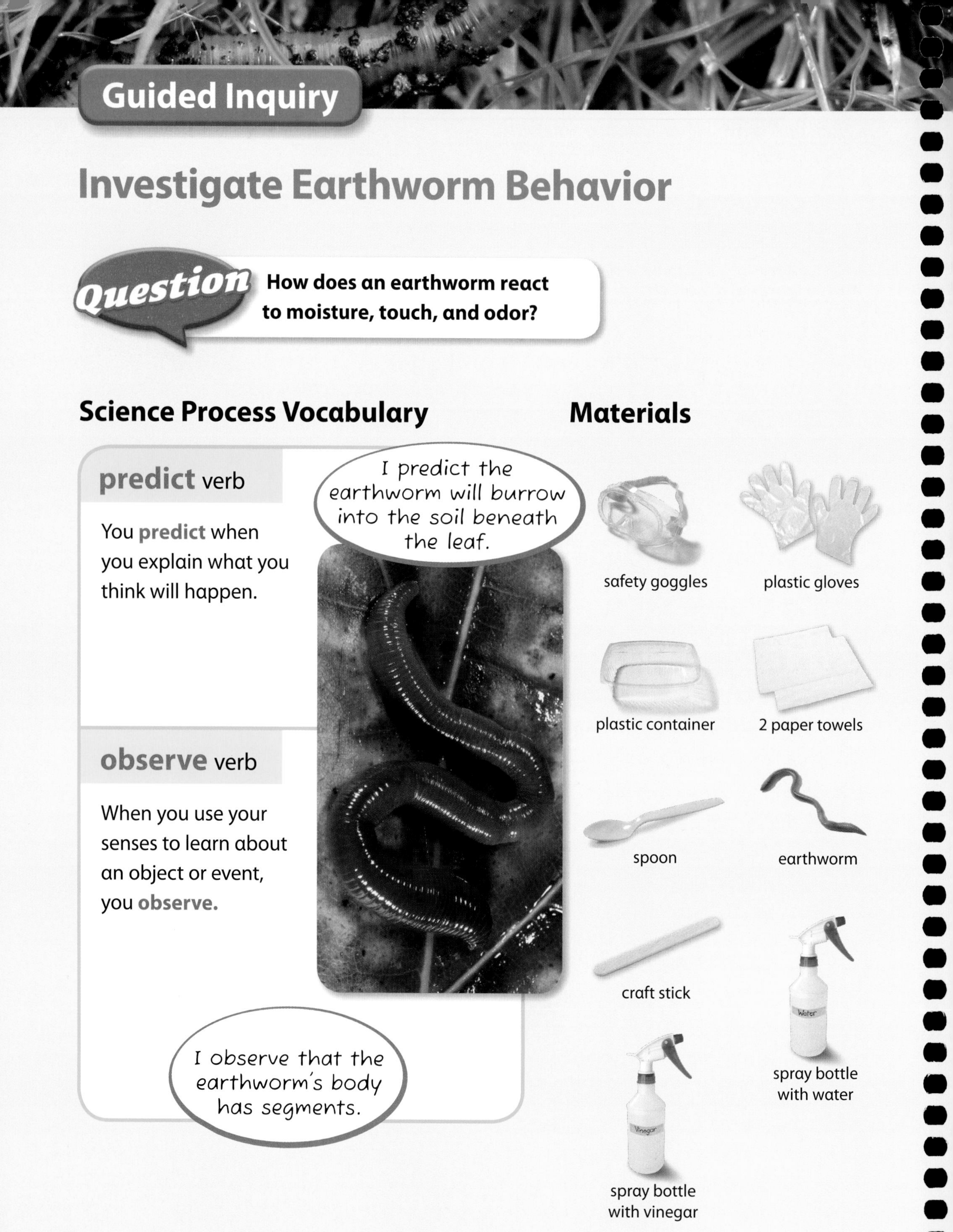

Guided Inquiry

Investigate Earthworm Behavior

Question **How does an earthworm react to moisture, touch, and odor?**

Science Process Vocabulary

predict verb

You **predict** when you explain what you think will happen.

observe verb

When you use your senses to learn about an object or event, you **observe.**

Materials

- safety goggles
- plastic gloves
- plastic container
- 2 paper towels
- spoon
- earthworm
- craft stick
- spray bottle with water
- spray bottle with vinegar

What to Do

1. Put on your safety goggles and plastic gloves. Fold the paper towels and place them side-by-side in the bottom of the plastic container.

2. Use the spoon to gently place the earthworm on the paper towel. **Predict** what will happen if you touch the earthworm near its head with a craft stick. Record your prediction in your science notebook. Use a craft stick to gently touch the earthworm near its head. **Observe** the earthworm's behavior. Record your observations.

Look for the "swollen ring" around the earthworm. The end near the ring is the earthworm's head.

What to Do, continued

3. Predict how the earthworm will react if you touch the middle of its body. Record your prediction. Then gently touch the middle of the earthworm with the craft stick. Record your observations.

4. Use the spoon to carefully remove the earthworm from the container and set it aside. Decide with your group whether you will observe the earthworm's reaction to moisture or to odor.

5. If you choose to observe the earthworm's reaction to moisture, use the spray bottle to moisten only 1 paper towel. Predict where the earthworm will move if you place it between the paper towels. Record your prediction. Then place the earthworm between the paper towels and observe its behavior. Record your observations.

6. If you choose to observe the earthworm's reaction to odor, use the spray bottle to moisten 1 paper towel with water. Then moisten the other paper towel with vinegar solution. Predict where the earthworm will move if you place it between the paper towels. Record your prediction. Then place the earthworm between the paper towels. Observe its behavior and record your observations.

Record

Write in your science notebook. Use a table like this one.

my SCIENCE notebook

Earthworm Behavior

	Reaction to Touch (Head)	Reaction to Touch (Middle)	Reaction to ______
Prediction			
Observation			

Explain and Conclude

1. Did your results support your **predictions?** Explain.
2. **Compare** the way the earthworm reacted to being touched near its head and near its middle.
3. **Infer** how the earthworm's behaviors help it survive.

Think of Another Question

What else would you like to find out about earthworm behavior? How could you find an answer to this new question?

Earthworms react to changes in the environment, such as heavy rain.

Directed Inquiry

Investigate Depth Perception

Science Process Vocabulary

count verb

When you **count,** you tell the number of something.

conclude verb

You **conclude** when you use information, or data, from an investigation to come up with a decision or answer.

I will use my observations and data to form a conclusion.

Materials

plastic cup

penny

meterstick

What to Do

1. Sit across from a partner at a table. Put the meterstick on the table. Place the plastic cup about 60 cm in front of your partner.

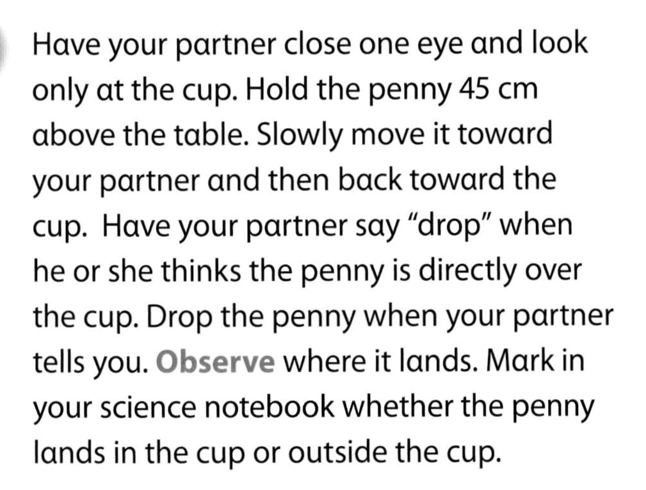

2. Have your partner close one eye and look only at the cup. Hold the penny 45 cm above the table. Slowly move it toward your partner and then back toward the cup. Have your partner say "drop" when he or she thinks the penny is directly over the cup. Drop the penny when your partner tells you. **Observe** where it lands. Mark in your science notebook whether the penny lands in the cup or outside the cup.

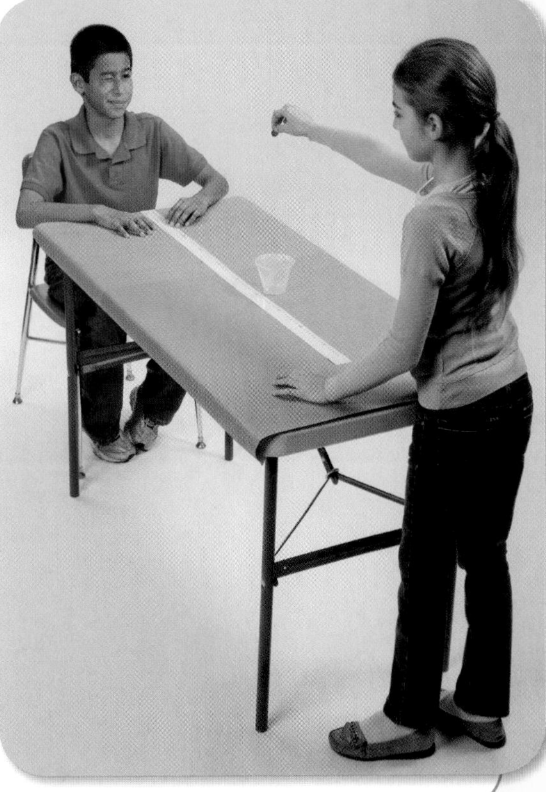

What to Do, continued

3. Repeat step 2 four more times. Have your partner close the same eye each time.

4. Repeat steps 2 and 3, but this time have your partner open both eyes.

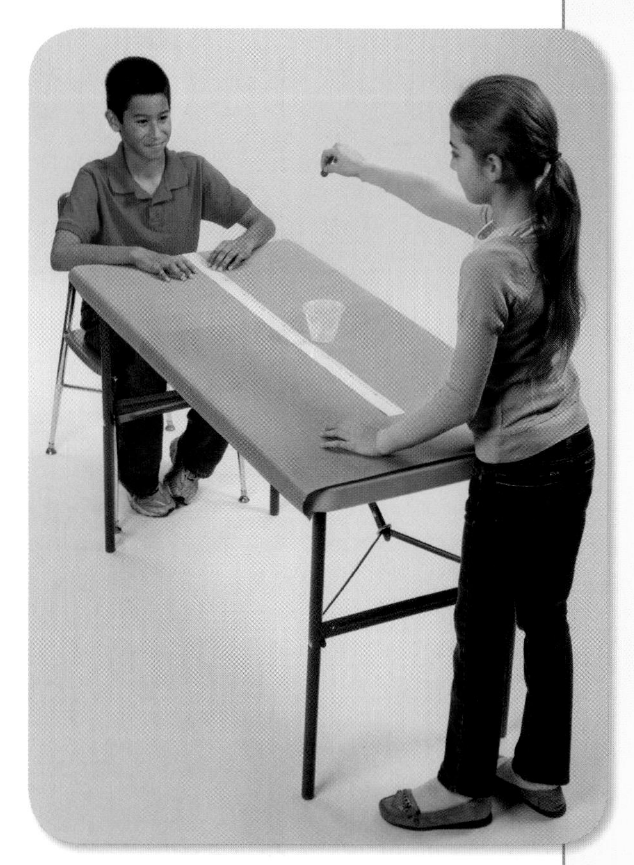

5. Switch roles with your partner and repeat steps 1–4.

Record

Write in your science notebook.
Use a table like this one.

my SCIENCE notebook

Penny Observations

	Penny Landed in Cup	Penny Landed Outside of Cup
1 eye open, trial 1		
1 eye open, trial 2		

Explain and Conclude

1. **Count** the number of times the penny landed in the cup. Did the penny land in the cup more often when one eye was open or when both eyes were open?
2. What can you **conclude** about how you use your eyes to judge depth?
3. Use the results of the **investigation** to **infer** why people need to be able to judge how far away objects are.

Think of Another Question

What else would you like to find out about how you use your eyes to judge depth? How could you find an answer to this new question?

The lion uses its eyes to judge how far away its prey is.

Guided Inquiry

Investigate Lungs and Carbon Dioxide

Question **How does exercise affect the amount of carbon dioxide in your breath?**

Science Process Vocabulary

experiment noun

In an **experiment,** you change only one variable, measure or observe another variable, and control other variables so they stay the same.

data noun

The observations and information that you collect and record in an investigation are **data.**

I will use the stopwatch to collect data.

Materials

safety goggles

3 cups of BTB solution

3 index cards

straw

stopwatch

Do an Experiment

Write your plan in your science notebook.

Make a Hypothesis

In this investigation, you will blow through a straw into a cup of BTB solution. This solution changes color when it is exposed to carbon dioxide. How will exercise affect the amount of time you must blow through the straw to change the color of the BTB solution?
Write your **hypothesis.**

Identify, Manipulate, and Control Variables

Which variable will you change?
Which variable will you observe or measure?
Which variables will you keep the same?

What to Do

1. Put on your safety goggles. Label an index card **No Exercise** and place it in front of the first cup of BTB solution. **Observe** the color of the BTB solution. Record the color in your science notebook.

No Exercise

2. Place a straw in the BTB solution. Have a partner observe the color of the solution as you blow gently into the straw. Your partner should use the stopwatch to **measure** how long you blow before he or she observes a change in the color of the solution. Record your **data** in your science notebook.

Be careful not to breathe in any liquid. Only blow outward through the straw.

No Exercise

What to Do, continued

3. Decide how many seconds you will run in place. Write the number of seconds on the index card and in your science notebook. Place the index card in front of the second cup of BTB solution. Place a straw in the cup. Record the color of the solution.

30 Seconds

4. Run in place for your chosen amount of time. Have your partner time your exercise.

5. Immediately after you have finished exercising, blow into the straw until your partner observes a change in color of the solution. Record how long you must blow before the change is noticeable.

6. Choose a different amount of time to exercise. Then repeat steps 3–5 for the third cup of BTB solution. Record your data in your science notebook.

Record

Write in your science notebook.
Use a table like this one.

my SCIENCE notebook

Exercise and BTB Solution

Time of Exercise (s)	Beginning Color of Solution	Blowing Time Needed to Change Color of the Solution (s)	Color of Solution After Change
No exercise (0 s)			

Explain and Conclude

1. Did your result support your **hypothesis?** Explain.
2. What happened to the color of the BTB solution when you blew through the straw into the solution?
3. Did it take more or less time to change the color of the solution after exercising? What can you **conclude** about how exercise affects the amount of carbon dioxide you breathe out from your lungs?

Think of Another Question

What else would you like to find out about lungs and exercise? How could you find an answer to this new question?

You use your lungs to breathe in the oxygen you need.

Do Your Own Investigation

Choose one of these questions, or make up one of your own to do your investigation.

- What happens to a plant cell when you add salt water to it?
- How can you use petroleum jelly and an index card to find out which place at school has the most particles in the air?
- Does bread mold grow best in wet or dry conditions?
- Which plant will grow best in salt water, rye grass or English ivy?
- What happens to your reaction time when you practice catching a ruler?

Science Process Vocabulary

question noun

You ask a **question** to find out about something.

How will the ivy react to salt water? I can investigate to answer this question.

investigate verb

You **investigate** when you make a plan and carry out the plan to answer a question.

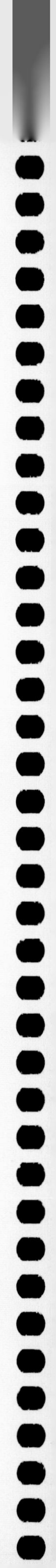

Open Inquiry Checklist

Here is a checklist you can use when you investigate.

- ☐ Choose a **question** or make up one of your own.
- ☐ Gather the materials you will use.
- ☐ If needed, make a **hypothesis** or a **prediction.**
- ☐ If needed, identify, manipulate, and control **variables.**
- ☐ Make a **plan** for your **investigation.**
- ☐ Carry out your plan.
- ☐ Collect and record **data. Analyze** your data.
- ☐ Explain and **share** your results.
- ☐ Tell what you **conclude.**
- ☐ Think of another question.

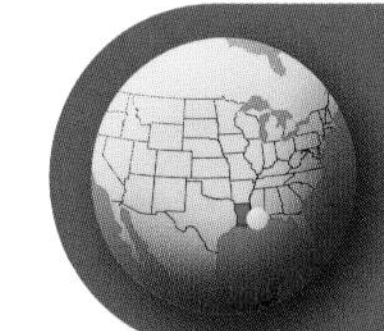

Bayou Sauvage National Wildlife Refuge, New Orleans, Louisiana

Some plants can tolerate salt water in the environment.

Write
Like a Scientist

Write About an Investigation

Plants and Salt Water

The following pages show how one student, James, wrote about an investigation. James wanted to grow some plants along the street at his home, but he knew that in winter, road crews applied salt to the road when it snowed. James wanted to make sure he used plants that could tolerate the salt. He decided to do an investigation. Here is what he thought about to get started:

Salt-tolerant Plants

Salt marsh plants grow well in water that contains salt.

- James knew that the salt that was applied to the road got into the soil where he wanted to grow his plants. Any plant that grew along the road would have to be able to live with the salt. He knew from visiting a salt marsh that some plants could live in salty environments.
- He wanted to use common plants that he could find easily. He decided to test rye grass and English ivy. He wanted his other materials to be simple and safe.
- James decided to do an investigation to find out how well rye grass and English ivy grow when they are watered with salt water. He would observe how the plants grow in the salty environment.

Model

Question

Which plant will grow best in salt water, rye grass or English ivy?

Make sure your question states clearly what you want to find out.

Materials

2 cups with soil and English ivy plants

2 cups with soil and rye grass plants

tape

salt

spoon

water

graduated cylinder

plastic cup

Gather all of your materials before you begin your investigation.

my SCIENCE notebook Your Investigation

Now it's your turn to do your investigation and write about it. Write about the following checklist items in your science notebook.

- [ ] **Choose a question or make up one of your own.**
- [ ] **Gather the materials you will use.**

Model

My Hypothesis

If I water English ivy and rye grass plants with both salt water and plain water, then the rye grass plants will grow well in both cups. The English ivy will grow well when watered with plain water, but will not grow well when watered with salt water.

James hypothesized that the rye grass would tolerate salt water because it looked similar to a salt marsh plant.

Your Investigation

my SCIENCE notebook

☐ **If needed, make a hypothesis or prediction.**

Write your hypothesis or prediction in your science notebook.

Model

Variable I Will Change

I will add salt water to 1 cup with English ivy and 1 cup with rye grass. I will add plain water without salt to another cup with English ivy and another cup with rye grass.

Variable I Will Observe or Measure

I will observe the color and shape of each plant to see which plants are healthy.

Variables I Will Keep the Same

I will keep all other variables the same. The plants will be in cups of the same size and will receive the same amount of water and light each day. I will grow them in the same temperature.

Answer these three questions:
1. Which variable will I change?
2. Which variable will I observe or measure?
3. Which variables will I keep the same?

my SCIENCE notebook

Your Investigation

☐ **If needed, identify, manipulate, and control variables.**

Write about the variables for your investigation.

Model

My Plan

1. Label the cups **Rye Grass Salt, Rye Grass No Salt, English Ivy Salt,** and **English Ivy No Salt.**
2. Place the 4 cups in a sunny spot.
3. Make a saltwater solution by mixing 1 level spoonful of salt with 50 mL of water in a cup. Label the cup **Salt Solution.**
4. Add 2 spoonfuls of the saltwater solution to the **Rye Grass Salt** cup. Repeat for the **English Ivy Salt** cup.
5. Add 2 spoonfuls of plain water into the cups without salt.
6. Repeat steps 4 and 5 and observe the plants each day for 1 week.

James chose a mixture of salt and water that he thought would be similar to conditions of the roads in winter.

Write detailed steps. Another student should be able to follow the steps without asking questions.

my SCIENCE notebook

Your Investigation

☐ **Make a plan for your investigation.**

Write the steps for your plan.

Model

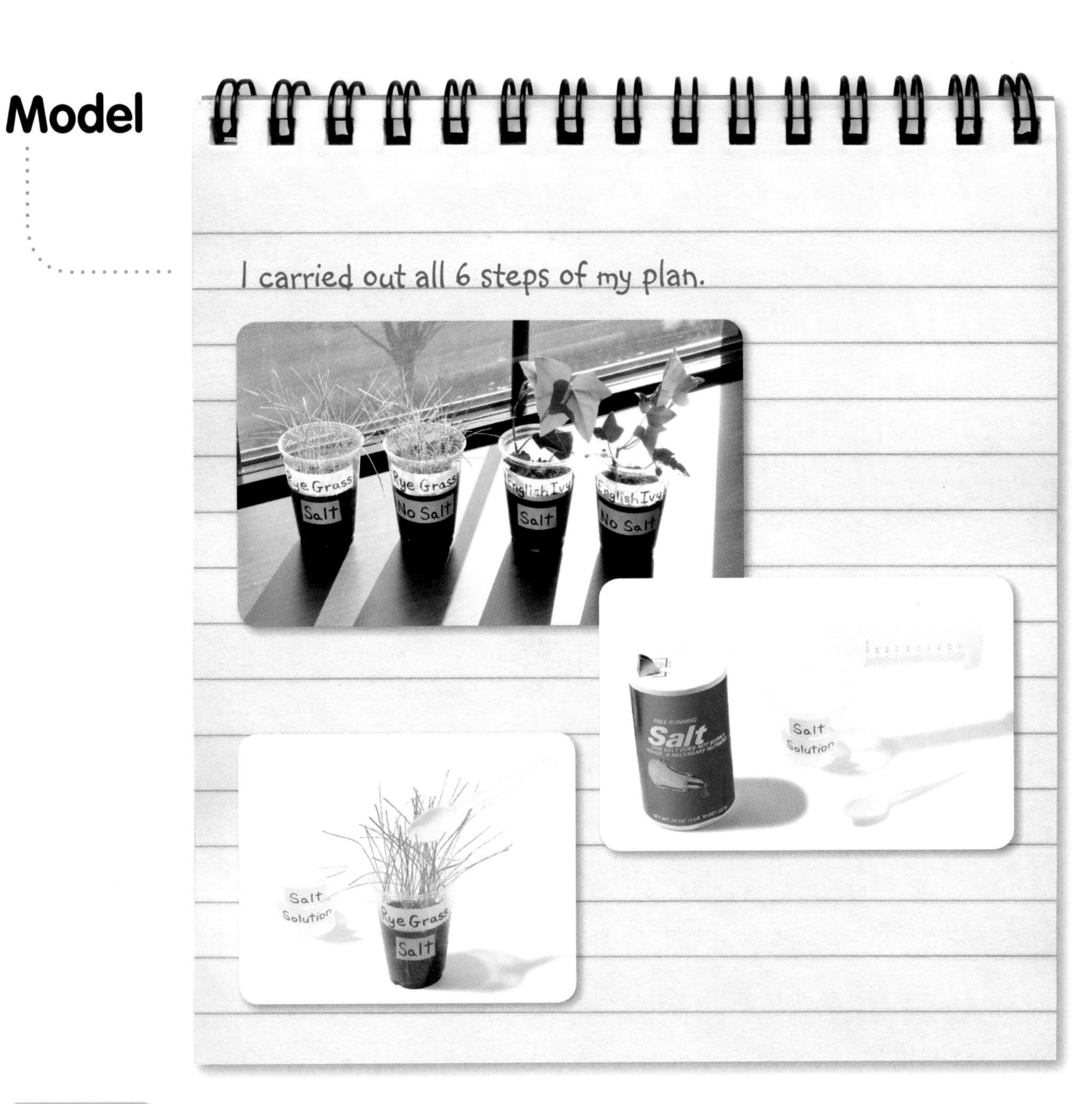

my SCIENCE notebook Your Investigation

☐ Carry out your plan.

Be sure to follow your plan carefully.

Model

Data (My Observations)

Observations of Plant Growth

Day	Rye grass: Salt	Rye grass: No salt	English ivy: Salt	English ivy: No salt
1	Green and standing up straight	Green and standing up straight	Green and healthy	Green and healthy
2	No change	No change	No change	No change
4	Turning yellow and brown	No change	No change	No change
5	Wilting	No change	No change	No change

My Analysis

The rye grass with no salt stayed healthy. The rye grass that had salt water added wilted. The English ivy that had no salt stayed healthy. The English ivy that had salt water added also was healthy.

Explain what happened based on the data you collected.

my SCIENCE notebook

Your Investigation

☐ **Collect and record data. Analyze your data.**

Collect and record your data, and then write your analysis.

Model

Scientists often share results so others can find out what was learned.

How I Shared My Results

I made a presentation to the class. First I explained the materials I used. Next I shared the results from the investigation. Then I told how my results show which plants can tolerate salt.

Tell what you conclude and what evidence you have for your conclusion.

My Conclusion

The results did not support my hypothesis. The English ivy grew well in the salt water, so it can tolerate salt. The rye grass did not grow well in the salt water, so rye grass does not tolerate salt. I conclude that some plants, like English ivy, can stay healthy in saltwater environments but others cannot.

Investigations often lead to new questions for Inquiry.

Another Question

I wonder how other variables, such as fertilizer, affect how plants grow. Can you give a plant too much fertilizer?

my SCIENCE notebook

Your Investigation

- ☐ Explain and share your results.
- ☐ Tell what you conclude.
- ☐ Think of another question.

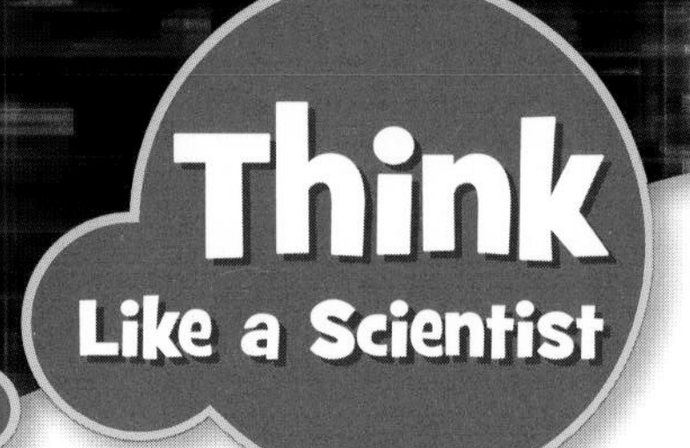

How Scientists Work

Scientific Investigations

Scientists do scientific investigations to solve problems and answer questions about the natural world. When they investigate, scientists follow an inquiry plan that uses process skills to gather, organize, analyze, and communicate information.

Scientific Methods

Inquiry plans that scientists use to answer questions are called scientific methods. The plan they use is not always the same. All investigations use scientific methods, but in many investigations, the methods are used in different ways. The order or number of steps might change, depending on the question.

Scientists often use tools to make better observations.

Planning Investigations When scientists plan an investigation, they may choose to do an experiment involving variables. Or they may do other kinds of investigations that involve observing living things, objects, or events without changing variables.

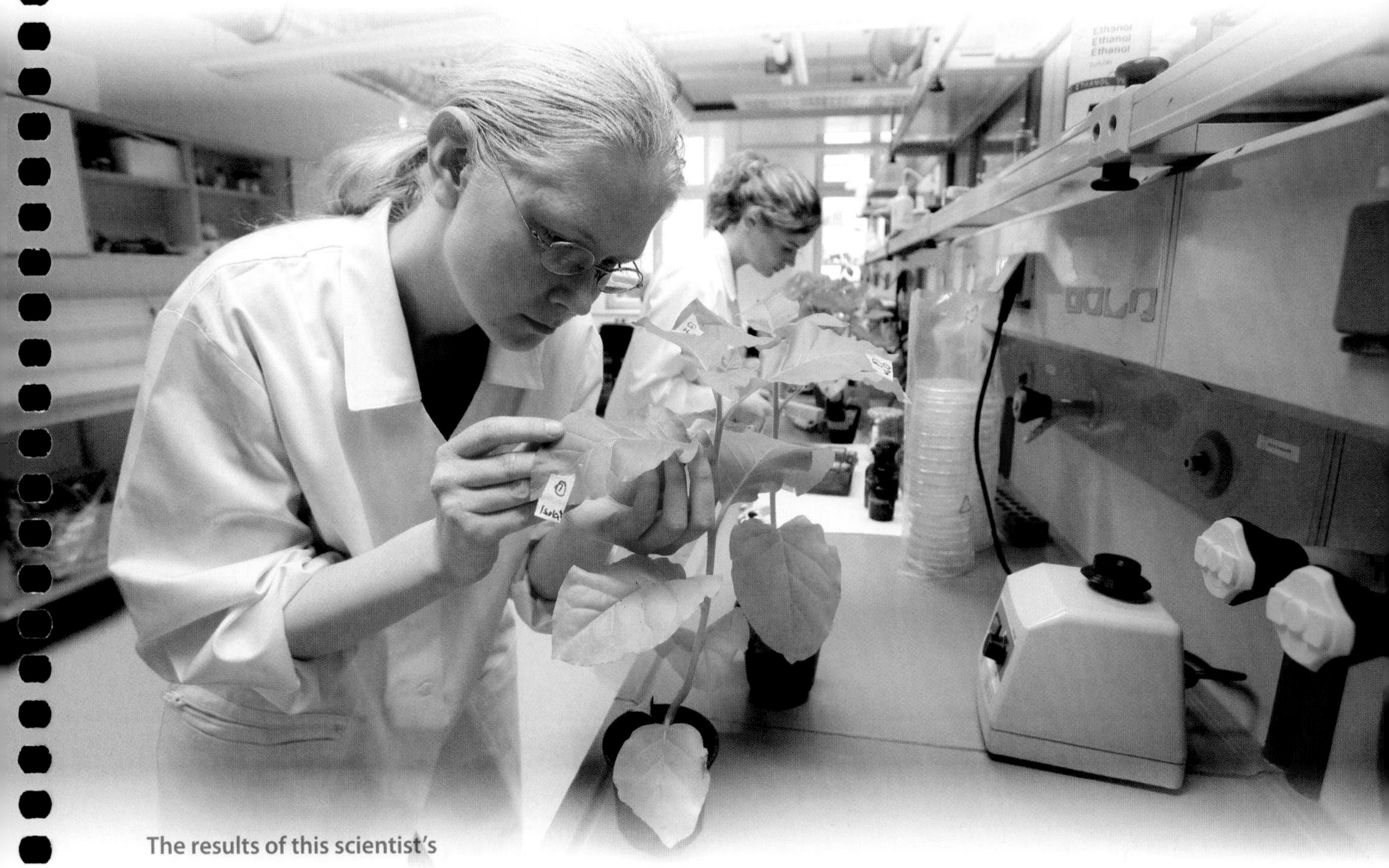

The results of this scientist's experiment may or may not support her hypothesis.

Experiments An experiment is one type of investigation. Scientists begin an experiment by forming a hypothesis, or a possible answer to their question. Then they perform a test to find out if the hypothesis is correct.

In an experiment, scientists think about the variables that can affect the outcome of the test. They choose one variable to manipulate, or change. The other variables stay the same. This way, scientists know that the results are caused by the variable they changed.

Observation Investigations Not all investigations are experiments in which variables are manipulated and controlled. Many investigations involve carefully planned observations of objects, events, or living things over time. Sometimes scientists build and observe models to learn more about processes in real life.

Collecting and Analyzing Data

Investigations can follow different scientific methods, but scientists always must collect and record data. The data might include measurements, written descriptions, or drawings. Scientists analyze their data to see if the data can help them answer their question. They draw conclusions about what the data means. Sometimes, the results of the investigation answer the question. Other times, the scientists must design a different kind of investigation to answer the question.

SUMMARIZE

What Did You Find Out?

1. Why do scientists conduct investigations?

2. What is the difference between an experiment and other types of scientific investigation?

 Plan an Investigation

Scientists think that a certain kind of butterfly in your area visits one kind of plant (plant A) more often than it visits another kind of plant (plant B). Design an investigation that could test which type of plant the butterflies visit more often.

1. Think of the question you want to answer.
2. Decide what type of investigation you will conduct.
3. Write the materials and scientific methods you will use.
4. Explain why you think your methods are the best way to answer the question. Is your investigation an experiment or an observation investigation?

CHAPTER 1

Science in a Snap! Model Earth's Movement Around the Sun

Have a partner make a fist to represent the sun. Use your right hand to be a **model** of Earth. Then make a "thumbs down" sign. Tilt your hand and thumb slightly to the right to model the tilt of Earth on its axis. Model Earth's movement around the sun. Keep your "Earth" hand tilted and move it slowly around your partner's "sun" hand. Have a partner say "Summer" when your model shows Earth's position when it is summer in the Northern Hemisphere.

Science in a Snap! Identify Objects in the Solar System

Write clues on an index card about each of the following objects in the solar system: the sun, an inner planet, an outer planet, a moon, an asteroid, and a comet. Show the cards to a partner and have your partner try to identify the object described on each card. Switch roles and try to guess the objects described on your partner's clue cards. Which clues were most helpful in identifying each object?

CHAPTER 3

Science in a Snap! Compare the Hardness of Pencils

The dark core of a pencil contains a mixture of clay and a mineral called graphite. If the mixture contains more graphite than clay, the pencil is soft and makes a dark mark. If the mixture has more clay, the pencil is hard and makes a lighter mark. Pencils are often marked with numbers or letters to show how hard or soft they are. Collect three pencils with different numbers or letters. Draw a line with each pencil. **Compare** the hardness of each pencil and the darkness of the marks. Which pencil is hardest?

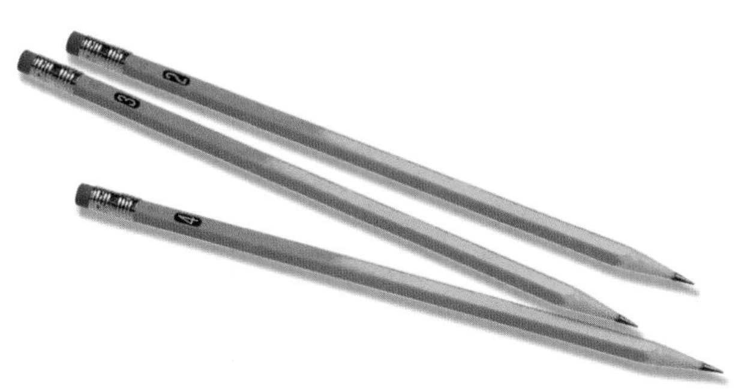

CHAPTER 4

Science in a Snap! Estimate Water Use

The chart below shows how much water the average person uses during everyday activities. **Estimate** how many times you do each activity in one day. Use the information in the chart to calculate how many liters of water you use each day. **Compare** your daily water use with a partner. What are some ways you can use less water each day?

Average Daily Water Use

Activity	Water use
Drinking water	200 mL per glass
Flushing toilets	15 L per flush
Taking a shower	19 L per minute
Washing dishes	40 L
Washing hands	1 L
Brushing teeth	4 L

Science in a Snap! Make a Rain Gauge

Use a ruler to mark 10 cm on a piece of tape. Stick the tape to a tall, clear container. You have made a rain gauge. Place your rain gauge outside in a flat, open area. You may need to use modeling clay or tape to hold your rain gauge in place. After a rainfall, **measure** the amount of rain in the rain gauge to the nearest centimeter and record your **data.** Be sure to empty the rain gauge after each rainfall. How much rain fell in 1 week? What else did you **observe** about the weather during the week?

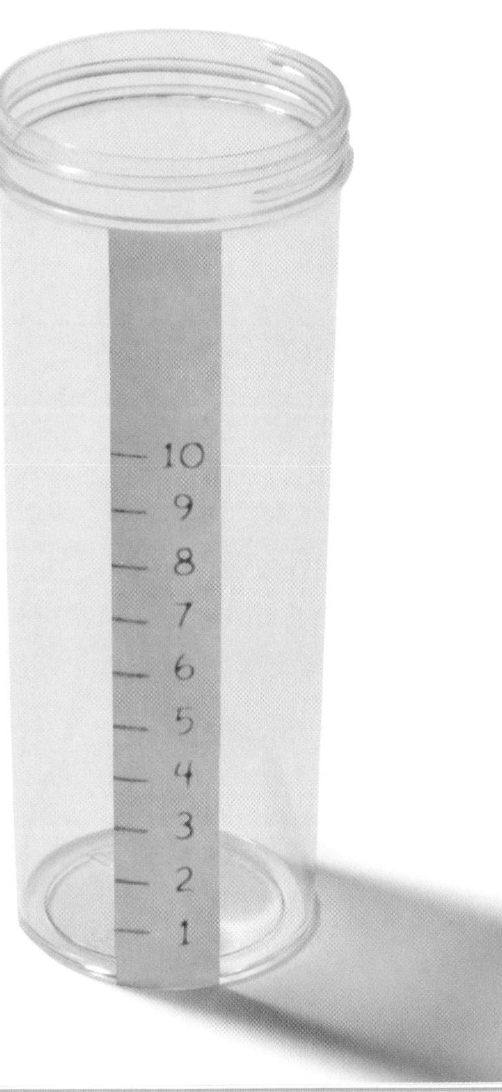

Explore Activity

Investigate Shadows and Time

Question How can you use shadows caused by sunlight to tell time?

Science Process Vocabulary

observe verb

When you **observe,** you use your senses to learn about an object or event.

estimate verb

When you **estimate,** you tell what you think about how much or how many.

I can use my observations of the sundial to estimate the time.

Materials

safety goggles

scissors

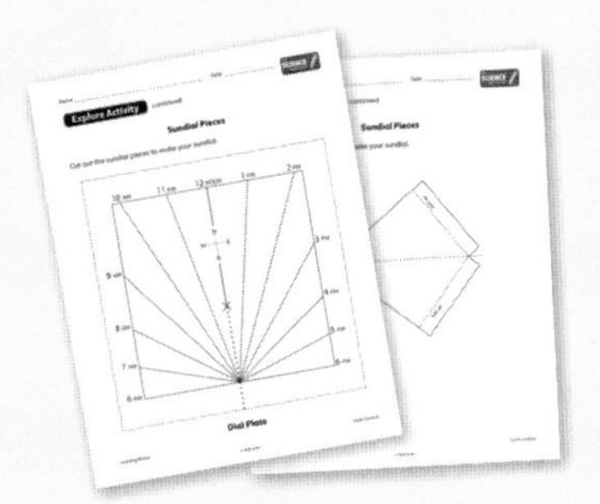

Sundial Pieces

tape

construction paper

compass

clock

ruler

What to Do

1. Put on your safety goggles. Cut out the sundial pieces. Cut along the dotted line to the **X** in the center of the dial plate.

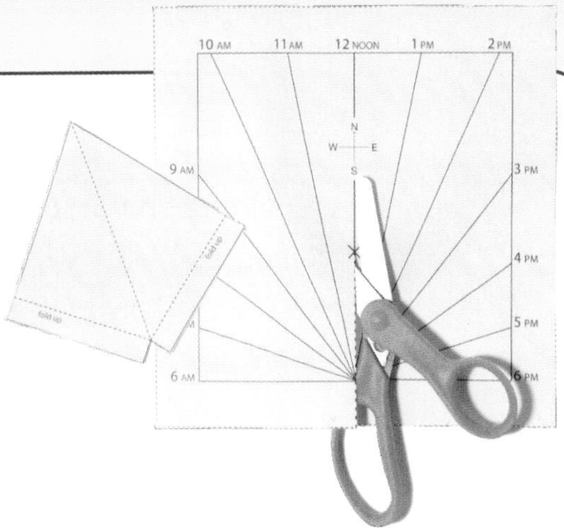

2. Fold the smaller sundial piece in half along the dotted line and fold the flaps up. Tape the sides together, as shown. This piece will be the gnomon, or pointer, in your sundial.

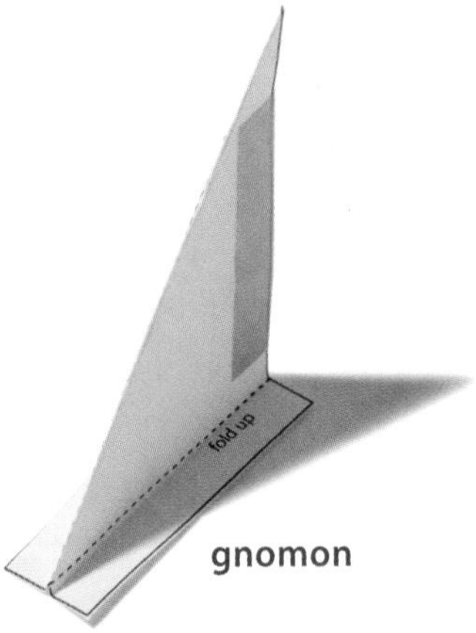

gnomon

3. Slide the gnomon into the slit in the dial plate. The tall side of the gnomon should be in the center of the dial plate. Tape the edges of the dial plate to the construction paper.

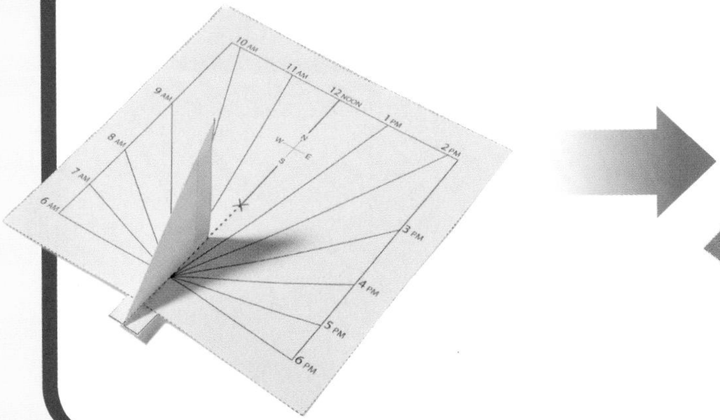

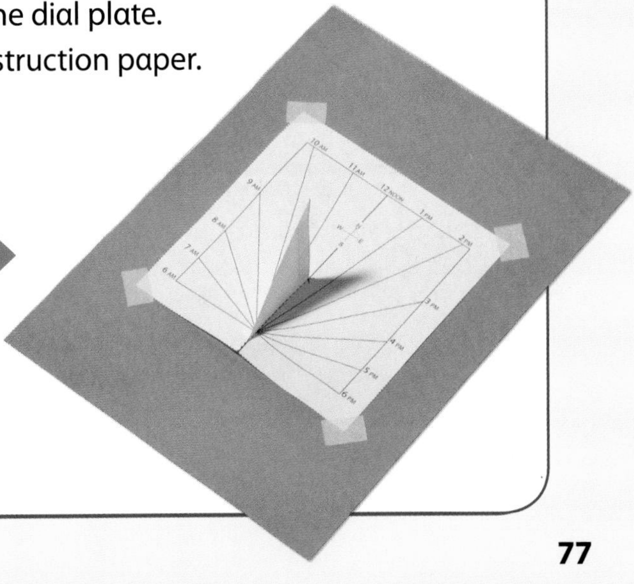

What to Do, continued

4. Place your sundial on a flat surface in a sunny spot. Use the compass to make sure that the 12 noon mark faces north.

The needle of the compass shows which direction is north.

5. **Observe** where the shadow of the gnomon falls on the dial plate. Draw your sundial and shadow in your science notebook. **Measure** the length of the shadow. Label the length of the shadow. Also record the position of the sun relative to the shadow.

Use the hour lines on the dial plate to estimate the time.

6. **Estimate** the time of day that is shown on your sundial. Record your estimate. Then check the actual time on a clock and record it in your science notebook. **Predict** what will happen to the shadow throughout the day. Record your prediction.

7. Repeat steps 5 and 6 every hour throughout the day.

Record

Write and draw in your science notebook. Use a table like this one.

my SCIENCE notebook

Observations of Sundial

Time	Draw your Sundial	Length of Shadow (cm)	Relative Position of Sun	Estimated Time	Actual Time
Start					
After 1 hour					

Explain and Conclude

1. **Compare** positions and length of the shadows on the sundial throughout the day.
2. When was the shadow longest? When was it shortest?
3. How did the position of the sun in the sky change throughout the day? How is the sun's position related to the position and length of shadows?

Forbidden City, Beijing, China

This ancient sundial is from the Ming Dynasty, which lasted from 1368 to 1644.

Directed Inquiry

Investigate the Phases of the Moon

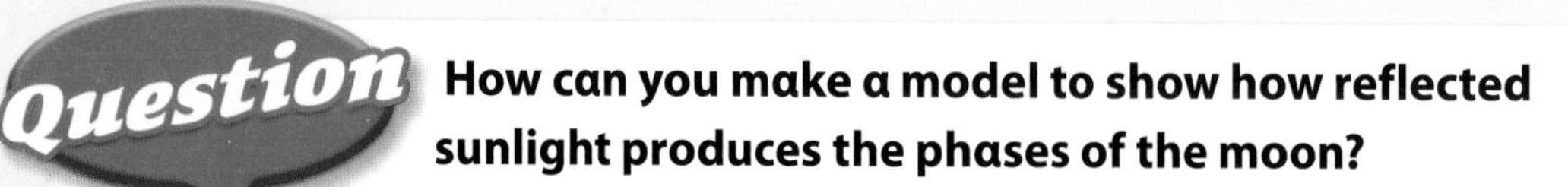

How can you make a model to show how reflected sunlight produces the phases of the moon?

Science Process Vocabulary

model noun

You can use a **model** to show why the phases of the moon occur.

I can use a model to find a pattern in the phases of the moon.

plan noun

When you make a **plan,** you list the materials you will need and the steps you will take.

I can make a plan to model what causes the phases of the moon.

Materials

lamp

ball

meterstick

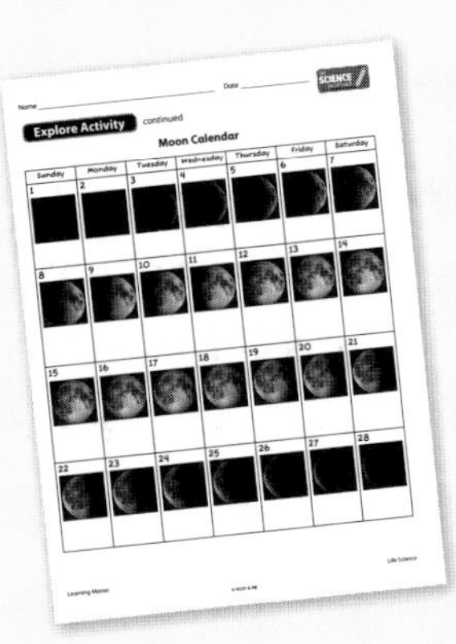

Moon Phases Calendar

What to Do

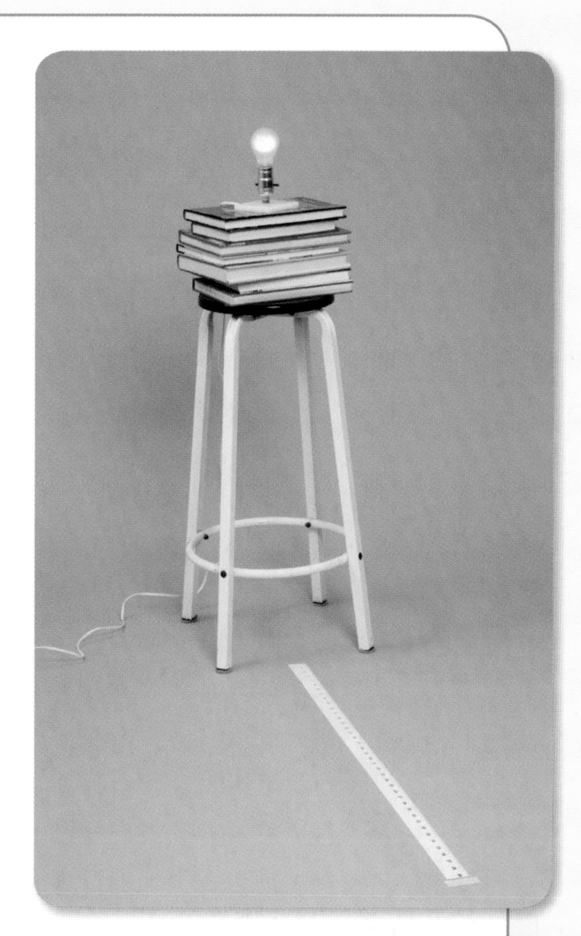

1. Your teacher will turn the lights off in the room. Place the lamp at eye level and turn it on.

2. Make a **model** of the sun-Earth-moon system. Measure 2 m from the lamp. Stand there and hold a ball. You should be facing the lamp. The lamp represents the sun. The ball represents the moon. Your head represents Earth.

3. Have a partner stand in front of you and behind the lamp. A second partner should stand about 2 m behind you. Raise the ball so that it is in front of you and slightly above your head.

What to Do, continued

4 Slowly rotate your body in a circle. As you rotate, **observe** the lighted portion of the ball as you turn in a complete circle. Have your partners also observe the light on the ball. Draw your observations and those of your partners in your science notebook.

5 Look at the Moon Phases Calendar. Notice the pattern of the lighted part of the moon during one month. Talk with your partners about how you can reproduce all the phases in the correct order using the ball and the light. Then try your **plan.**

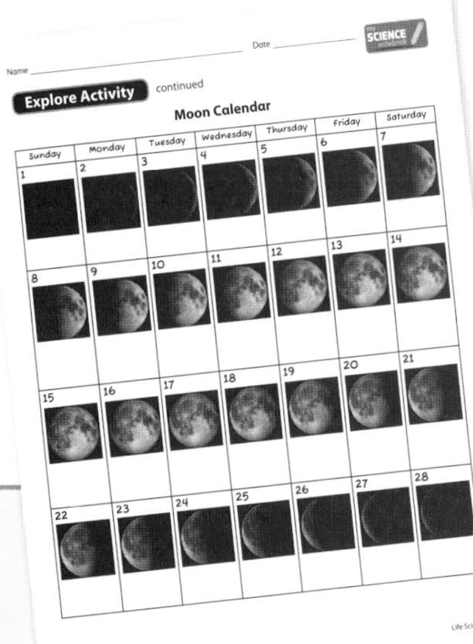

Name ________________ Date ________________ SCIENCE

Explore Activity continued

Moon Calendar

Sunday	Monday	Tuesday	Wednesday	Thursday	Friday	Saturday
1	2	3	4	5	6	7
8	9	10	11	12	13	14
15	16	17	18	19	20	21
22	23	24	25	26	27	28

Life Science

Learning Master © NGSP & HB

Record

Write and draw in your science notebook.
Use a table like this one.

my SCIENCE notebook

Observations of Ball

Where I Was	Me	Partner Behind the Light	Partner Behind Me
1 2			

Explain and Conclude

1. Based on the **observations** of your partners, how much of the moon **model** is lighted at any one time?
2. In step 5, did you move in a clockwise or in a counterclockwise direction to produce the phases of the moon in the correct order?
3. Use your observations and those of your partners to explain what causes moon phases.

Think of Another Question

What else would you like to find out about the phases of the moon? How could you find an answer to this new question?

Surface of the moon

Directed Inquiry

Investigate Earth and the Moon

Question **How can you model the distance between Earth and its moon?**

Science Process Vocabulary

model noun

You can use a **model** to show how something in real life works. Scientists often use models to study objects that are too big or too far away to study directly.

estimate verb

When you **estimate,** you tell what you think about how much or how many.

I estimate that the second ball is two times as large as the first ball.

Materials

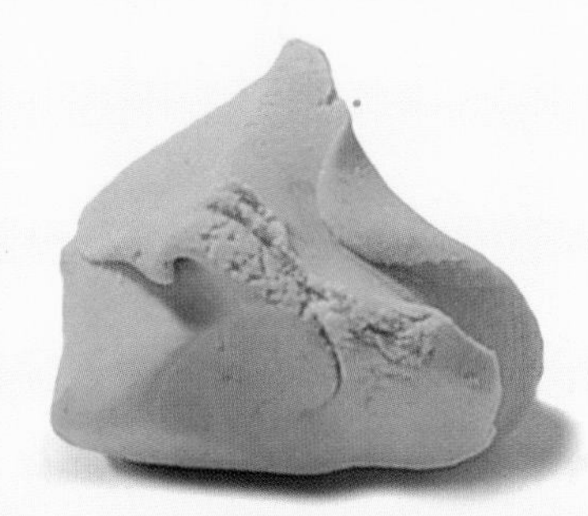

modeling clay

ruler

What to Do

1. Make 2 clay balls that represent Earth and its moon. **Estimate** how big you think Earth is compared to the moon. Make your **models** the sizes you think they should be.

2. Use the ruler to **measure** the distance across the middle of each ball. The distance across the middle of the ball is its diameter. Record your **data** in your science notebook.

3. Put the clay balls together to make 1 lump. Use 1 piece of clay to make a ball with a diameter of 4 cm. Use another piece of clay to make a ball with a diameter of 1 cm. Record the diameters of the clay balls. You have now made a model of Earth and its moon that represents the correct difference in size. A model that shows how sizes and distances compare is called a scale model.

What to Do, continued

4. Estimate how many centimeters apart you think the clay balls should be in your model. The distance in your model represents the distance between the moon and Earth in space. Record your estimate in your science notebook. Move the model Earth and moon to show what you think the distance should be.

5. In space, the average distance between the moon and Earth is about 30 times the diameter of Earth. Multiply the diameter of your model Earth by 30 to find out how many centimeters apart the clay balls should be in your model. Record the distance.

6. Move the model Earth and moon so that they are the correct distance apart. Draw your model. Add labels to your drawing to show the correct diameters of the model Earth and moon and the distance between them.

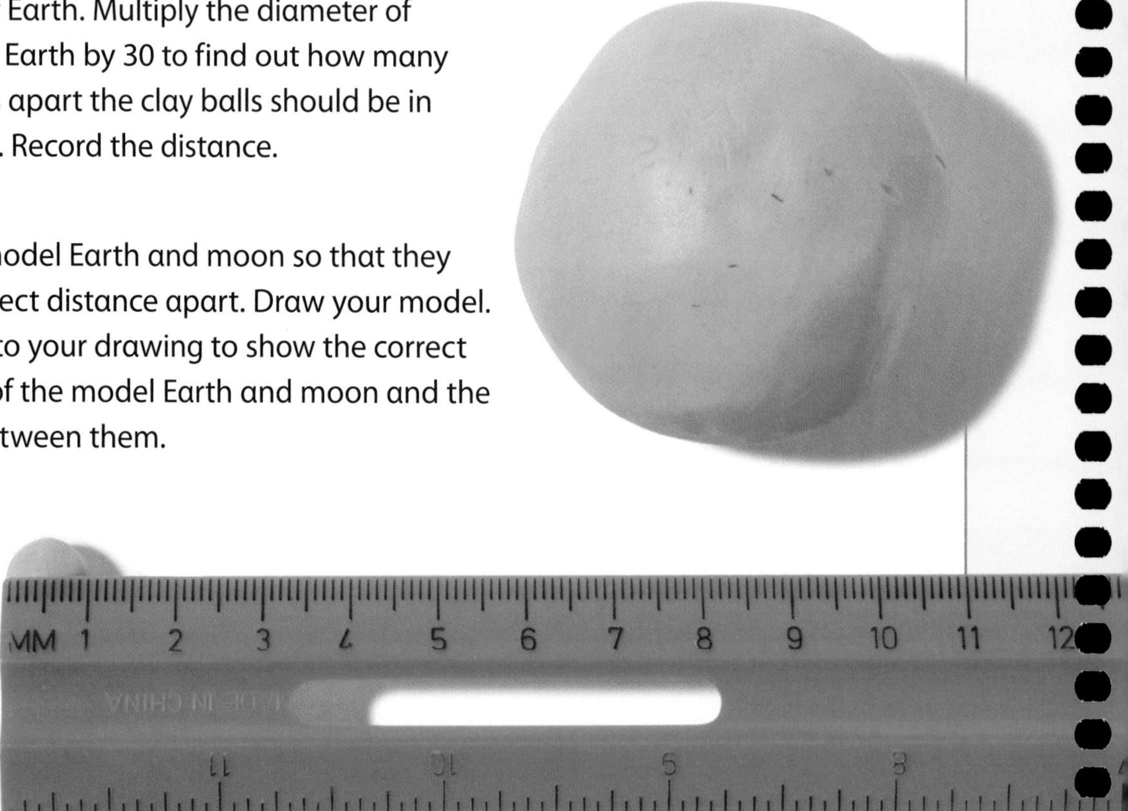

Record

Write and draw in your science notebook.
Use a table like this one.

my SCIENCE notebook

Models of Earth and Its Moon

	Diameter of Model Earth (cm)	Diameter of Model Moon (cm)	Distance Between Earth and Moon (cm)
Estimated model			
Scale model			

Model Earth and Moon

Explain and Conclude

1. **Compare** the diameters of your estimated **models** with your scale models.
2. What **data** do you need to make a scale model of Earth and the moon?
3. What makes models useful for learning about Earth and the moon?

Think of Another Question

What else would you like to find out about modeling Earth and the moon? How could you find an answer to this new question?

Scientists can observe Earth from the International Space Station.

Think
Like a Scientist

Math in Science

Interpreting and Comparing Large Numbers

When you read about the sun, moon, stars, and Earth, you may see numbers that represent very large amounts. Sometimes the amounts are so large that understanding them is difficult.

As you learn about the solar system, you can use a few techniques to understand how large and far apart everything is. When you come across very large numbers, you can think of those numbers in a language you understand.

The average distance of our moon from Earth is 384,403 km.

Interpreting Large Numbers

When you see a very large number, connect it to something you know. For example, Earth is about 150 million km from the sun. That is written 150,000,000 km. How can you understand how far that is?

One way to understand the distance from Earth to the sun is to use something more familiar—the speed of a car on a highway. The average car travels about 100 km per hour on the highway. At that speed it would take you 1,500,000 hours to get from Earth to the sun. That's about 171 years!

Here is another example. Earth's circumference (the distance around it) is about 40,000 km at the equator. You would have to travel around Earth about 3,750 times to travel the same distance you would travel from Earth to the sun. Does that help you understand how far away the sun is?

Earth and the sun

Comparing Large Numbers

Sometimes you may need to compare very large numbers. For example, if you wanted to compare the distance of Mercury and Venus from the sun, you might say that Mercury is about 58,000,000 km away while Venus is about 108,000,000 km away.

An easy way to compare the distances is to use a shortened form, like the one used in this chart.

Pay attention to the words *million kilometers* at the top of the right column. These words tell you that the measurements on the chart are in kilometers, and each number should be followed by six zeroes. Using the chart, you can easily see the relative distances of the planets from the sun. For example, it is easy to see that Venus is almost twice as far from the sun as Mercury.

Planet	Average Distance from the Sun (million kilometers)
Mercury	58
Venus	108
Earth	150
Mars	228
Jupiter	779
Saturn	1,434
Uranus	2,873
Neptune	4,495

Venus (left) and Mercury

SUMMARIZE

What Did You Find Out?

What techniques can you use to help you understand very large numbers?

How can a chart help you compare large numbers?

Practice
Comparing Numbers

The numbers below show the diameter of each planet rounded to the nearest thousand kilometers. Make a chart that uses a shortened form to compare the diameters of the planets more easily. Then answer the questions.

- Mercury: 5,000 km
- Venus: 12,000 km
- Earth: 13,000 km
- Mars: 7,000 km
- Jupiter: 143,000 km
- Saturn: 116,000 km
- Uranus: 47,000 km
- Neptune: 45,000 km

This illustration shows the relative sizes of the sun and planets. The relative distances between the sun and planets are not shown.

 How do the diameters in your shortened form differ from the actual diameters?

 Which planet has a diameter about half that of Earth's diameter?

 Which planet has the largest diameter?

Guided Inquiry

Investigate the Solar System

Question How can you make a scale model that shows the sizes of the sun and planets in our solar system?

Science Process Vocabulary

model noun

Scientists use **models** to learn more about objects that are too big or too far away to observe directly.

The model shows how the planets' sizes compare to each other in real life.

measure verb

When you **measure,** you find out how much or how many.

Materials

ruler

masking tape

scissors

yarn

3 sheets of construction paper

What to Do

1. You will make a scale **model** that shows the sizes of the sun and planets in our solar system. The chart shows the actual diameters of the sun and planets. Diameter is the distance across the center. The Scale Diameter columns of the chart give two choices of what the diameters of the sun and planets could be in your scale model.

Part of the Solar System	Diameter (km)	Scale A Diameter (cm)	Scale B Diameter (cm)
Sun	1,391,000	109.0	54.5
Mercury	4,866	0.4	0.2
Venus	12,106	0.9	0.45
Earth	12,756	1.0	0.5
Mars	6,760	0.5	0.25
Jupiter	142,984	11.0	5.5
Saturn	116,438	9.0	4.5
Uranus	46,940	4.0	2.0
Neptune	45,432	3.8	1.9

2. Choose which scale you will use to find the diameters of your model sun and planets: Scale A or Scale B. Then find the scale diameter for the sun. Use the metric ruler to **measure** a strip of tape that is the same length as the scale diameter of the sun. Place the tape on the floor. Measure another strip with the same length and place it on the floor so the strips make the shape of a plus sign.

What to Do, continued

3. Cut a piece of yarn long enough to make a circle around the strips of tape. This circle will represent the sun in your model.

4. For Earth, draw a plus sign on a sheet of construction paper. The lines that make up the plus sign should be 1 cm long if you chose Scale A or 0.5 cm long if you chose Scale B. Draw a circle around the plus sign. Cut out the circle. Label the circle **Earth.**

5. Repeat step 4 for each planet. Use the Scale Diameter column you chose in the chart to tell you how big to make the plus sign for each planet.

6. Place the planets on the floor in the correct order from the sun. **Observe** your planets and the sun. Draw your model in your science notebook. Be sure to label the sun and planets.

Record

Draw a picture of your solar system model in your science notebook.

Scale Model of the Solar System

Explain and Conclude

1. Which planet is largest? Which is smallest? How does your **model** help show this?
2. Describe Earth's position in the solar system. Use your model to help you.
3. Why is it important for a model of the sizes of the sun and planets to be a scale model?

Think of Another Question

What else would you like to find out about making models of the solar system? How could you find an answer to this new question?

Jupiter

Directed Inquiry

Investigate Soil

Question **What minerals and other materials can you observe in soil?**

Science Process Vocabulary

observe verb

You can **observe** soil to see its different parts.

compare verb

When you **compare,** you tell how objects are alike and different.

Materials

quartz

feldspar

hornblende

hand lens

safety goggles

spoon

soil

paper

forceps

microscope

What to Do

1. Use the hand lens to **observe** the 3 minerals: quartz, feldspar, and hornblende. Bits of these minerals are commonly found in soil. Record the properties of each mineral in your science notebook.

2. Put on your safety goggles. Place a spoonful of soil on a sheet of paper. Spread out the soil. Observe the soil. Draw what you see.

What to Do, continued

3. Observe the soil with the microscope. Look for bits of plants, rocks and minerals, insects, seeds, or any other materials. Draw what you see. Label any materials you can identify.

4. Observe the soil with the hand lens. Use the forceps to separate the different materials in the soil. Sort the materials into groups such as twigs, seeds, or small pieces of rocks and minerals. Draw some of the materials in each group. Label those you can identify.

5. Observe the pieces of rocks and minerals. **Compare** them to the minerals you observed in step 1. Draw and label them.

Record

Write and draw in your science notebook.
Use tables like these.

my SCIENCE notebook

Minerals

Mineral	Observations
Quartz	
Feldspar	
Hornblende	

Soil

Step 2: With Eyes Only	Step 3: With Microscope
Step 4: With Hand lens	Step 5: Minerals

Explain and Conclude

1. How many different kinds of materials did you **observe** in the soil? What materials were you able to identify?
2. Which minerals did you identify in the soil?
3. Was there more of one kind of mineral than others in the soil? If so, which one?

Think of Another Question

What else would you like to find out about the minerals and other materials in soil? How could you find an answer to this new question?

Guided Inquiry

Investigate Fossils

Question How can you use a model rock to infer about past environments?

Science Process Vocabulary

model noun

You can make and use a **model** to show how something in real life works.

infer verb

When you **infer,** you use what you know and what you observe to draw a conclusion.

I can observe this fossil and infer what the fish's environment was like.

Materials

trilobite fossil

crinoid fossil

hand lens

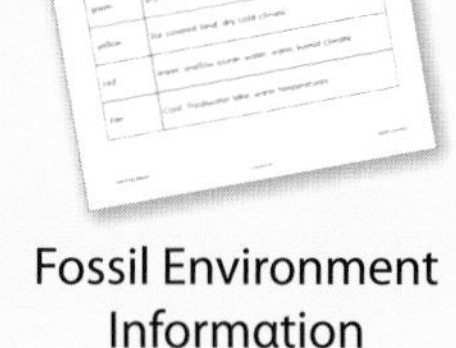

Fossil Environment Information

4 lumps of clay

4 small objects

knife

craft stick

toothpick

forceps

What to Do

1. Use the hand lens to **observe** the fossils. Record your observations in your science notebook. You will make a sedimentary rock **model,** like the rock that contained these fossils. Read the Fossil Environment Information.

2. Choose one color of clay and flatten it. The clay represents a layer of soil laid down millions of years ago. Choose a small object and press it into your clay layer. The small object represents an organism that died on the soil. Over time the soil formed rock. The organism became a fossil.

3. Use a different color of clay to add another layer over the rock. This clay layer represents soil that covered the first layer and then became rock. Choose another small object and press it into your second clay layer. Continue adding clay layers and objects until you have four layers on your model.

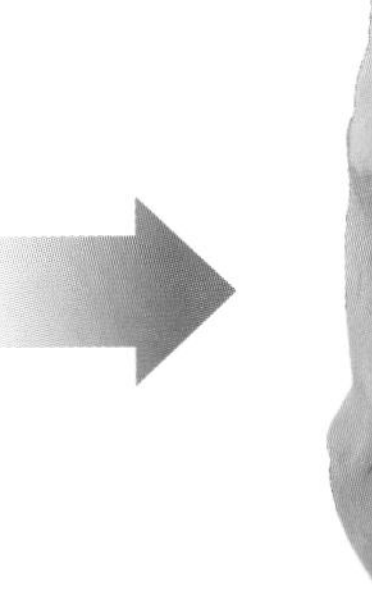

What to Do, continued

4. Exchange your model rock with another group. Use the knife to cut down through all the layers of the model rock you received. Draw the layers of the model rock in your science notebook.

5. Use the craft stick, toothpick, or forceps to remove the model fossils from the model rock. Record your fossil discoveries on your model rock drawing. Draw each model fossil in the layer in which it was found.

6. Use the Fossil Environment Information to find out what the environment was like when each of the model fossilized organisms lived. Record the information on your drawing of the model rock layers.

Record

Write and draw in your science notebook.
Use tables like these.

my SCIENCE notebook

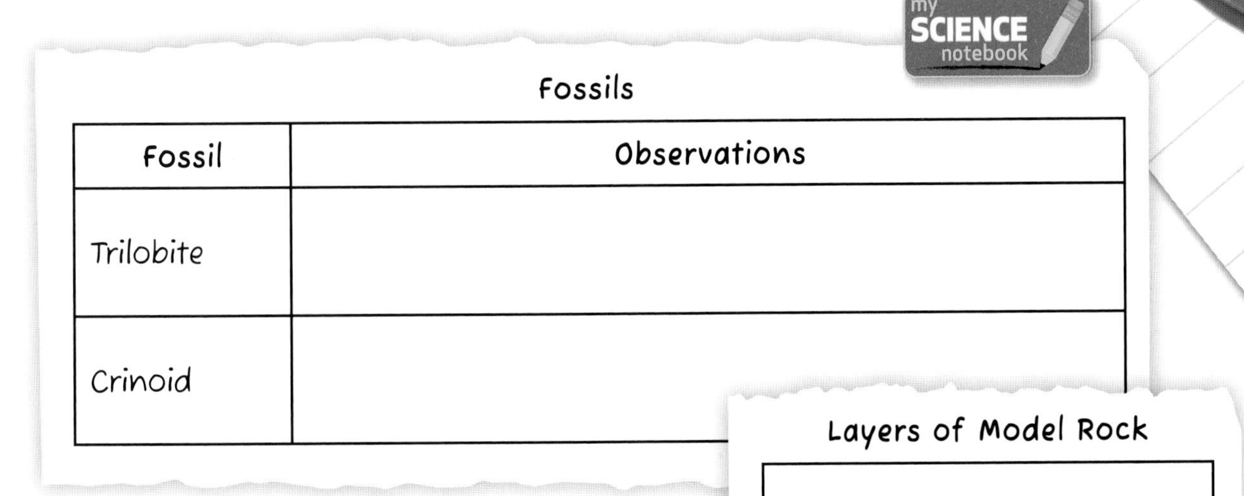

Fossils

Fossil	Observations
Trilobite	
Crinoid	

Layers of Model Rock

Explain and Conclude

1. Which layer of your **model** had the oldest model fossils? How do you know?
2. Fossils of trilobites and crinoids would belong in the red layer of rock. Describe their environment.
3. What can you **infer** about how the environment of the area where your fossil was found changed over the millions of years? Explain your reasoning.

Think of Another Question

What else would you like to find out about how to use fossils and rock to infer about Earth's past environments? How could you find an answer to this new question?

This fossilized *Diplomystus* fish is 34–56 million years old. It is related to herrings and sardines that live today.

Directed Inquiry

Investigate Recycling

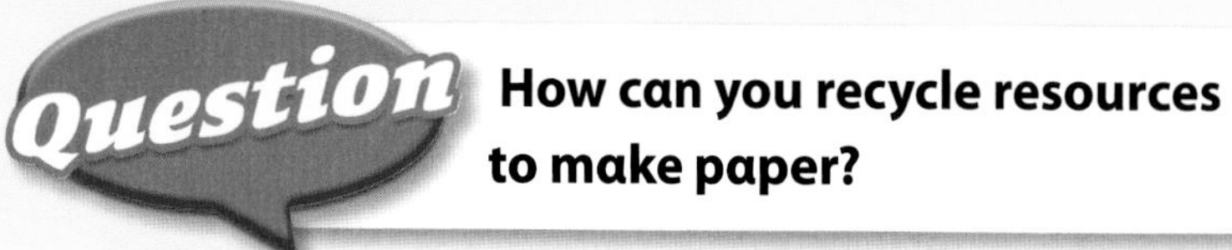

Question How can you recycle resources to make paper?

Science Process Vocabulary

measure verb

You can **measure** materials to get the correct amount.

compare verb

When you **compare** two or more objects, you tell how they are alike and different.

Materials

safety goggles

plastic container

water

measuring cup

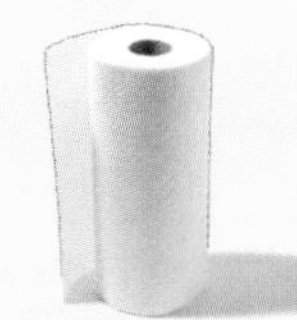

paper towels

colored paper pieces

measuring spoon

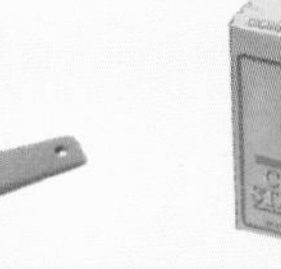

corn starch

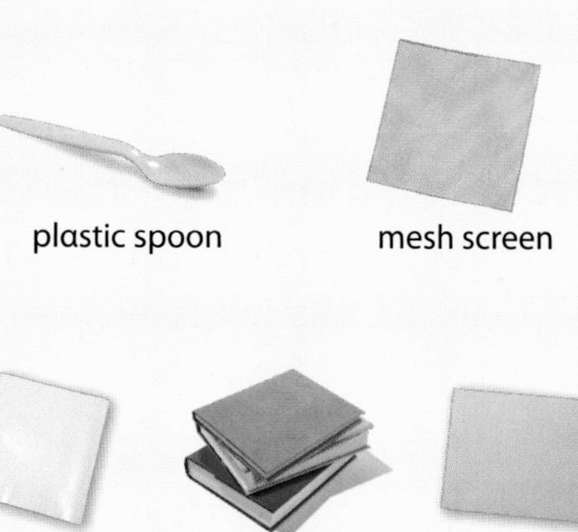

plastic spoon

mesh screen

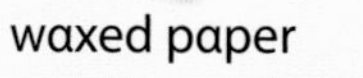

waxed paper

books

colored paper

What to Do

1. Put on your safety goggles. **Measure** and pour 2 cups of water into the plastic container.

2. Tear up 2 paper towels and place the pieces in the water. Add the used colored paper pieces to the container. Mash up the wet pieces with your fingers.

3. Measure and pour 2 measuring spoonfuls of cornstarch into the container. Stir the container's contents with the plastic spoon. The cornstarch and paper pieces will mix together to make a material called pulp.

What to Do, continued

4. Place the mesh screen on 3 paper towels. Pick up a handful of pulp and hold it over the container to allow excess water to drip from it. Then place the pulp on the screen. Continue removing the pulp from the container and placing it on the screen until all the pulp is on the screen. Spread the pulp over the screen evenly.

5. Place waxed paper over the pulp on the screen. Put the books on top of the waxed paper and gently press down on them to flatten the pulp. Remove the books and waxed paper.

6. Gently lift the screen and pulp from the wet paper towels and place them on 3 dry paper towels. Allow the screen to sit overnight.

7. When your paper is dry, **observe** its physical properties, such as color, texture, and thickness. Observe the properties of a regular sheet of colored paper. Write on each paper and tear each in half. Record your observations in your science notebook.

Record

Write in your science notebook.
Use a table like this one.

my SCIENCE notebook

Properties of Paper

Property	Recycled Paper	Regular Paper
Color		
Texture		
Thickness		

Explain and Conclude

1. **Compare** the properties of the recycled paper and the sheet of colored paper. How are they alike and different?
2. What are some advantages of using the recycled paper? What are some disadvantages?
3. How does recycling help to maintain Earth's natural resources?

Think of Another Question

What else would you like to find out about recycling? How could you find an answer to this new question?

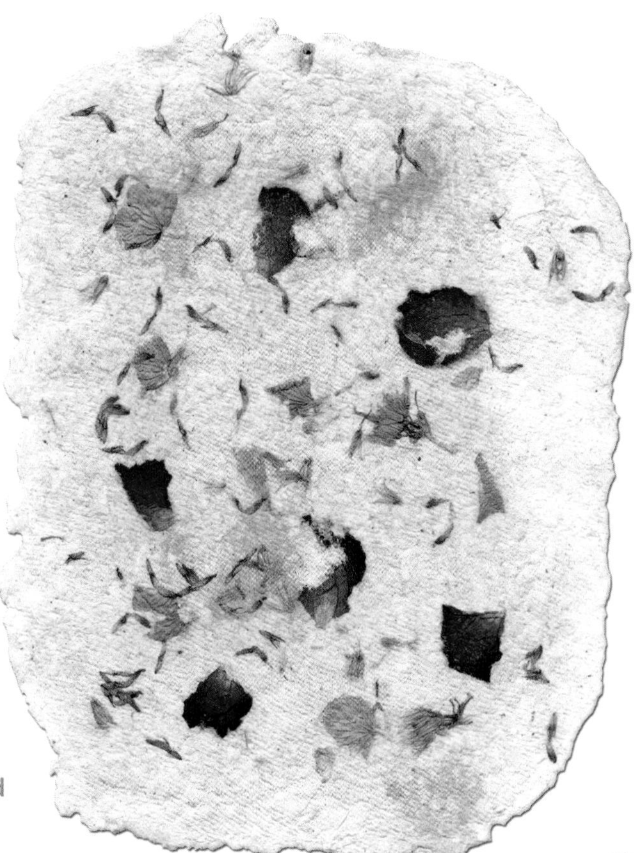

Leaves, flowers, herbs, and seeds can be added to recycled paper before it dries.

Guided Inquiry

Investigate Solar Energy

How can you use energy from the sun to make water cleaner?

Science Process Vocabulary

compare verb

When you **compare,** you tell how objects or events are alike and how they are different.

conclude verb

You **conclude** when you use information, or data, from an investigation to come up with a decision or answer.

I conclude that food coloring dissolves in water.

Materials

safety goggles

plastic basin

plastic cup

clay

water

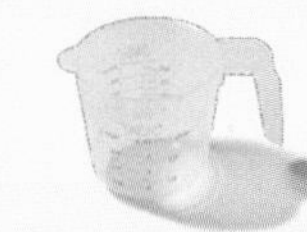
measuring cup

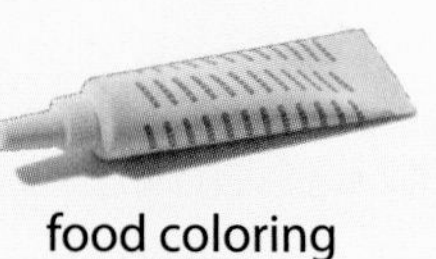
food coloring

sandy soil

drink mix

spoon

plastic wrap

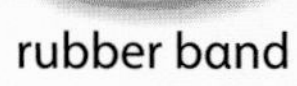
rubber band

rock

What to Do

1. Put on your safety goggles. Use the clay to stick the plastic cup to the center of the basin.

2. Use the measuring cup to pour 500 mL of water into the basin. Do not pour any water into the plastic cup.

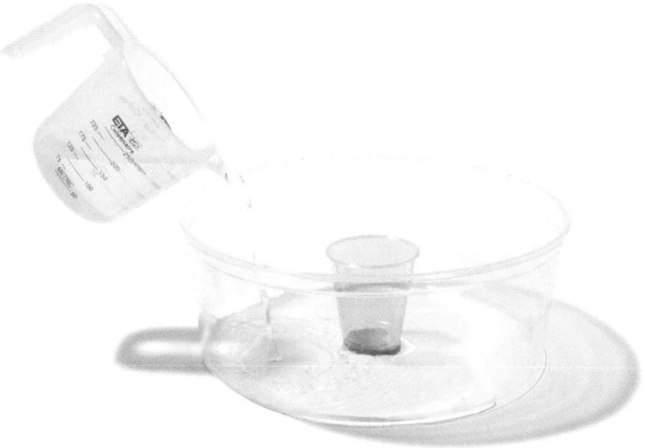

3. Choose a material to add to the water in the basin: soil, food coloring, or drink mix. Use the spoon to gently stir the water so that the material mixes with the water. **Observe** the plastic cup and the basin. Record your observations in your science notebook.

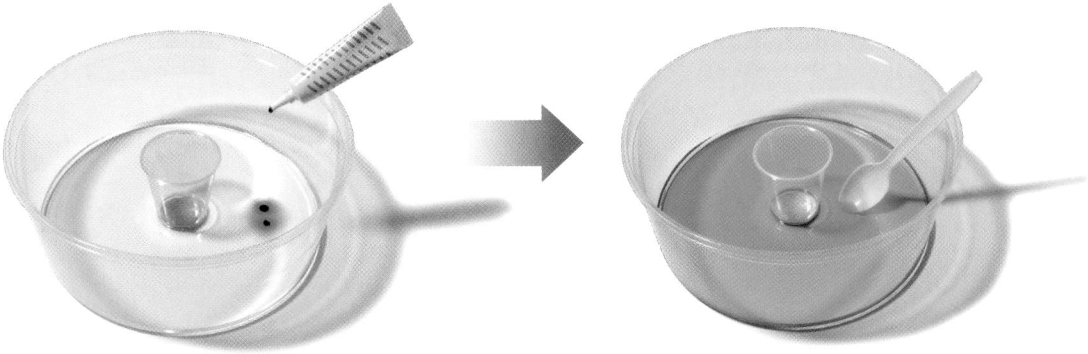

What to Do, continued

4 Cover the basin with plastic wrap. Use a rubber band to seal the plastic wrap tightly around the basin. Place the rock in the center of the plastic wrap, directly above the cup. Then put the basin in a sunny spot.

5 After 2 days, observe the plastic wrap and the water in the basin. Remove the plastic wrap and observe the plastic cup. Record your observations.

6 **Compare** your results with other groups. Were the results the same for groups that used drink mix, food coloring, and soil? Discuss ways that changing the design of the basin could improve your results.

Record

Write in your science notebook.
Use a table like this one.

my SCIENCE notebook

Water

	Material Added	Observations in Step 3	Observations in Step 5
Water in basin			
Cup			

Explain and Conclude

1. What caused the water to move from the basin to the cup?
2. **Compare** the water in the cup with the water in the basin. What can you **conclude** about how solar energy can be used to make water cleaner?
3. How could you change the design of your basin to clean the water more quickly? How could you change the design to clean greater amounts of water?

Think of Another Question

What else would you like to find out about how energy from the sun can be used to make water cleaner? How could you find an answer to this new question?

Kicking Horse River, Yoho National Park, British Columbia, Canada

Investigate the Water Cycle

Question **What happens to a model lake when sunlight shines on it?**

Science Process Vocabulary

observe verb

When you **observe,** you use your senses to learn about an object or event.

I can observe very small water droplets in the air.

predict verb

You **predict** when you use your observations and what you already know to say what will happen.

I predict that some water from the pond will evaporate.

Materials

safety goggles

plastic container

sandy soil

warm water

measuring cup

plastic wrap

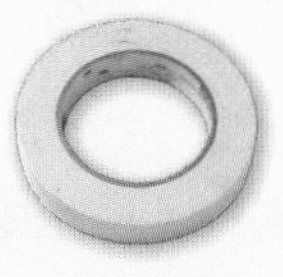

tape

What to Do

1. Put on your safety goggles. Add sandy soil to one end of the plastic container to make a **model** of a lake shore.

2. Pour 1 cup of warm water in the other end of the container. Cover the container with plastic wrap. Tape the plastic wrap to the container. The container is a model of a lake and the air around it. Set your model in a sunny spot.

What to Do, continued

3. **Observe** your model. Draw the model in your science notebook. Identify where water is located and whether the water is a solid, liquid, or gas. **Predict** what will happen to the water in 2 hours. Record your predictions and observations.

4. Observe the model after 2 hours. Has the water changed? If so, how? Record your observations. Predict what will happen to the water after 24 hours. Record your predictions.

5. Observe your model after 24 hours. Tap the plastic wrap and observe what happens. Has the water changed? If so, how? Record your observations.

Record

Write and draw in your science notebook.
Use a table like this one.

my SCIENCE notebook

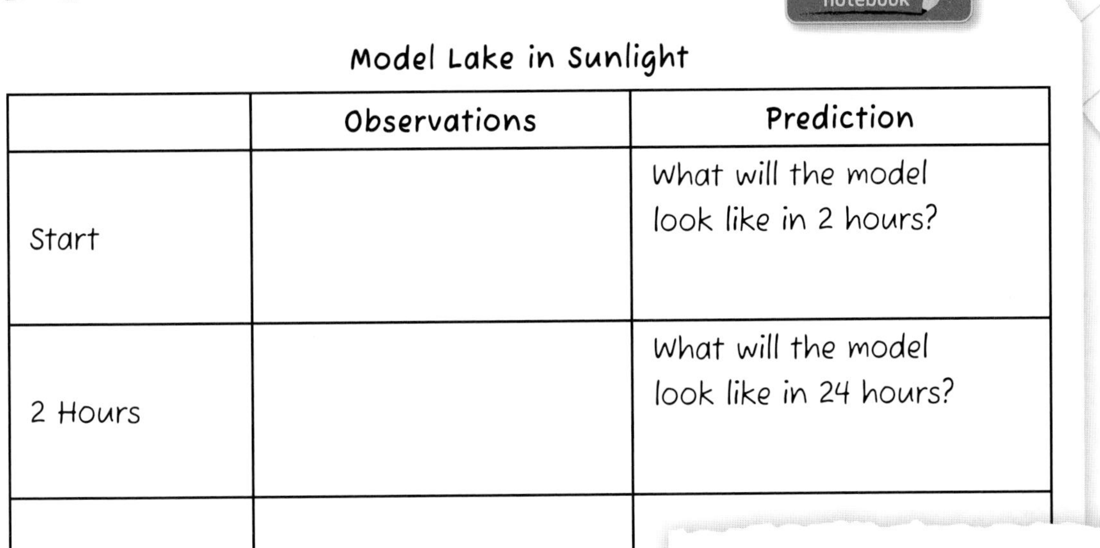

Model Lake in Sunlight

	Observations	Prediction
Start		What will the model look like in 2 hours?
2 Hours		What will the model look like in 24 hours?

Drawing of a Model Lake

Explain and Conclude

1. Did your results support your **predictions?** Explain.
2. How did water change during the investigation? What caused these changes?
3. Use what you **observed** about your **model** to explain what happens in the water cycle.

Think of Another Question

What else would you like to find out about the water cycle? How could you find an answer to this new question?

Water returns to Earth's surface as rain or other forms of precipitation.

Guided Inquiry

Investigate Weather

Question **How can you use weather tools to measure weather conditions?**

Science Process Vocabulary

measure verb

You **measure** when you find out how much or how many. Scientists use different tools to measure weather conditions.

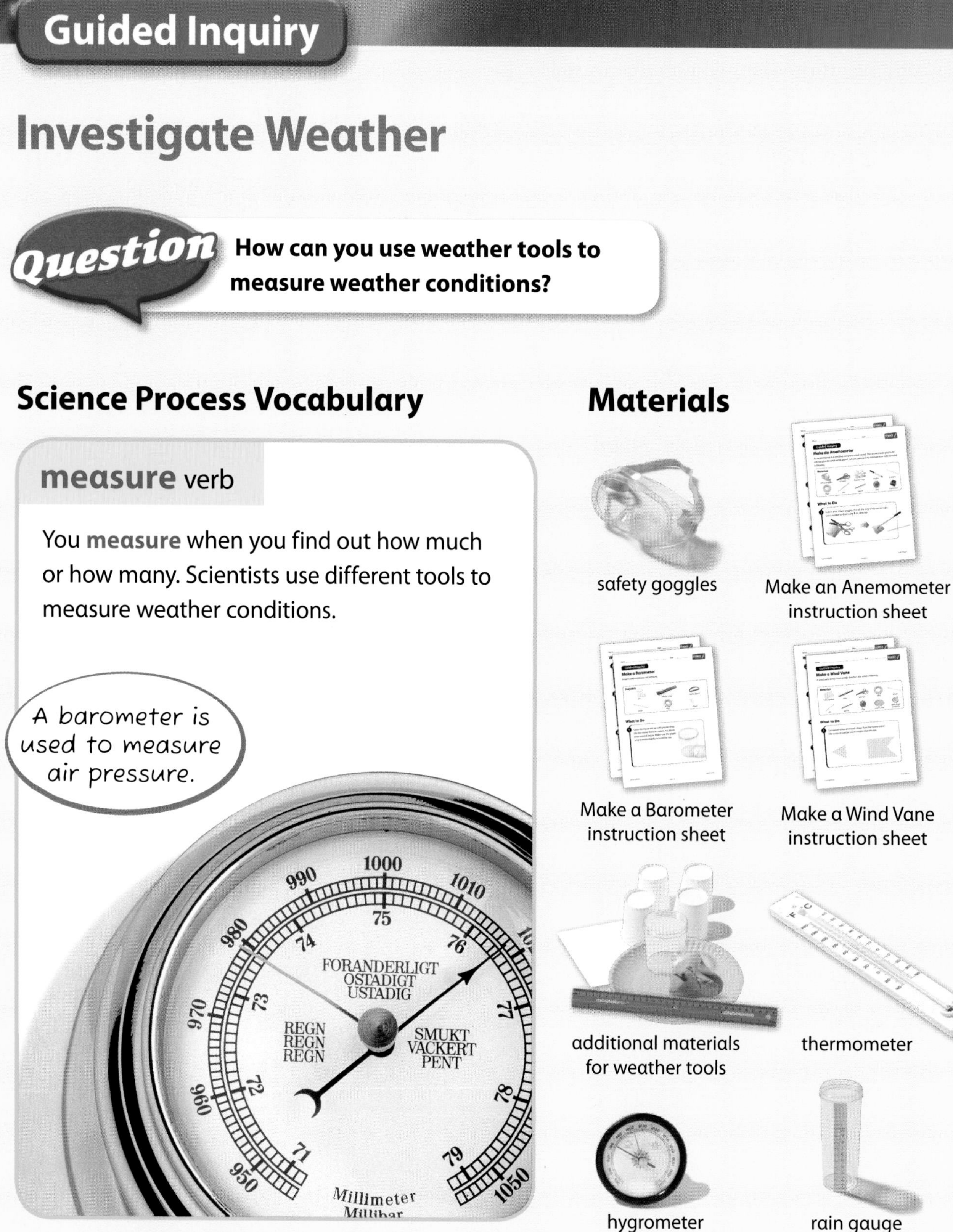

Materials

safety goggles

Make an Anemometer instruction sheet

Make a Barometer instruction sheet

Make a Wind Vane instruction sheet

additional materials for weather tools

thermometer

hygrometer

rain gauge

What to Do

1 You can make tools to help you **measure** weather conditions. With your group, choose the tool that you will build: an anemometer, a barometer, or a wind vane. Put on your safety goggles. Then follow the steps on the instruction sheet to build your weather tool.

2 All groups will put their weather tools outside to collect weather **data.** With your group, choose a location to place your weather tool. Choose a safe place where it will be easy to collect data.

3 Your teacher will also place the thermometer, hygrometer, and rain gauge outside for all groups to use.

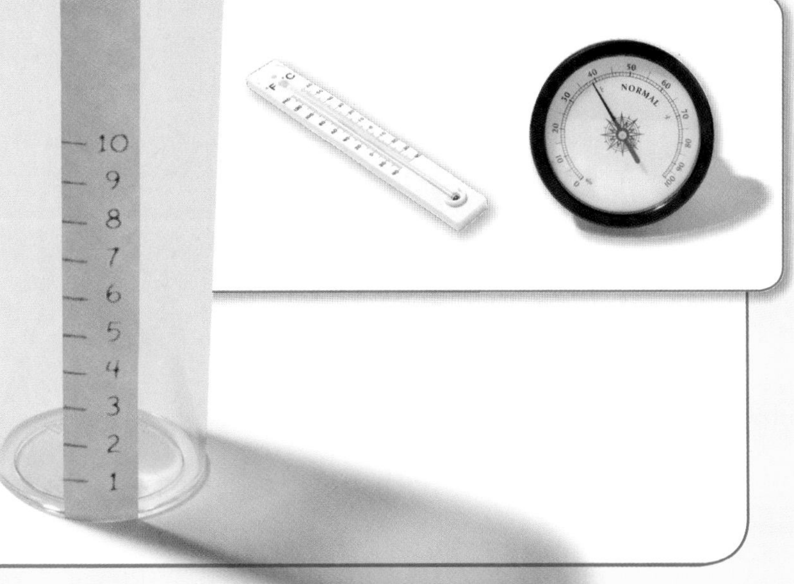

What to Do, continued

4 Record the data from all the weather tools every day for 3 weeks. Record your data in your science notebook.

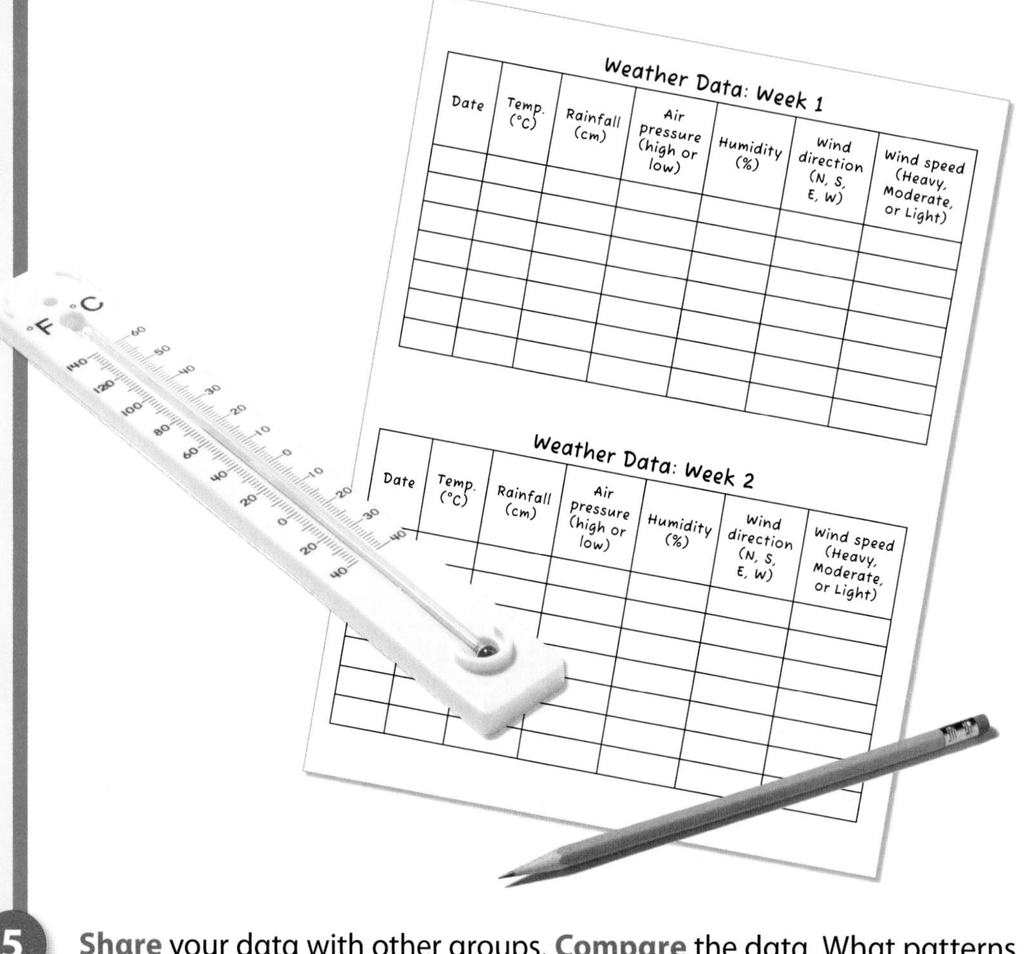

Weather Data: Week 1

Date	Temp. (°C)	Rainfall (cm)	Air pressure (high or low)	Humidity (%)	Wind direction (N, S, E, W)	Wind speed (Heavy, Moderate, or Light)

Weather Data: Week 2

Date	Temp. (°C)	Rainfall (cm)	Air pressure (high or low)	Humidity (%)	Wind direction (N, S, E, W)	Wind speed (Heavy, Moderate, or Light)

5 **Share** your data with other groups. **Compare** the data. What patterns do you see?

Record

Write in your science notebook. Use a table like this one. Fill in the columns for temperature, rainfall, humidity, and the weather condition you chose to measure.

Weather Data: Week 1

Date	Temp. (°C)	Rainfall (cm)	Air Pressure (high or low)	Humidity (%)	Wind Direction (N, S, E, W)	Wind Speed (heavy, moderate, or light)

Explain and Conclude

1. How did your weather tool help you to **measure** the weather conditions?
2. Did you see any patterns in your **data?** What patterns did you find in the class data?

Think of Another Question

What else would you like to find out about weather tools? How could you find an answer to this new question?

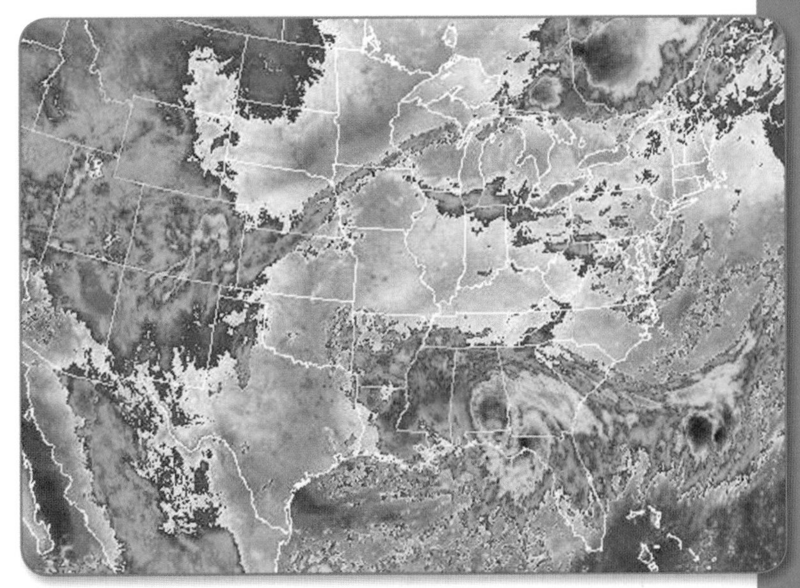

Radar is a tool that is used to track storms. Computers use data collected by radar to make maps like this.

Open Inquiry

Do Your Own Investigation

Choose one of these questions, or make up one of your own to do your investigation.

- How can you use a model to show how seasons change at different latitudes?
- How can you make a model of the shape of a spinning planet?
- How do table salt crystals compare to Epsom salt crystals?
- How does water filtered with gravel, sand, and charcoal compare to water that is unfiltered?
- How does the humidity of the air affect how quickly water vapor will condense on a can of ice water?

Science Process Vocabulary

hypothesis noun

When you make a **hypothesis,** you state a possible answer to a question that can be tested by an experiment.

If I add ice to warm water in a can, then water will condense on the outside of the can.

Open Inquiry Checklist

Here is a checklist you can use when you investigate.

- ☐ Choose a **question** or make up one of your own.
- ☐ Gather the materials you will use.
- ☐ If needed, make a **hypothesis** or a **prediction.**
- ☐ If needed, identify, manipulate, and control **variables.**
- ☐ Make a **plan** for your **investigation.**
- ☐ Carry out your plan.
- ☐ Collect and record **data. Analyze** your data.
- ☐ Explain and **share** your results.
- ☐ Tell what you **conclude.**
- ☐ Think of another question.

Dew forms on the grass when cool temperatures cause water vapor in the air to condense.

Write About an Investigation

Condensation

The following pages show how one student, Katie, wrote about an investigation. Katie knew that water vapor condenses when air cools. She wondered how the humidity of the air affects how quickly water vapor condenses. She decided to investigate the question. Here is what she thought about to get started:

- Katie wanted to do an investigation that compared how quickly water vapor condenses when different amounts of water vapor are in the air.
- She needed to use safe and simple materials. She would base her question and the steps she would take on the materials she could obtain.
- Katie decided to pour ice water in a metal can and measure the time needed for water vapor to condense.
- She would conduct multiple trials and use a hygrometer to measure the humidity in the air at the time of each trial.

Condensation

First, the air around the ice water cools.

The water molecules in the air slow down and move closer together.

The water molecules condense on the can's surface to form liquid water.

Model

Question

How does the humidity in the air affect how quickly water vapor will condense on a can of ice water?

Choose a question that can be answered using materials that are safe and easy to obtain.

Materials

graduated cylinder
200 mL ice water
thermometer
can
hygrometer
stopwatch

List exact amounts of materials when needed.

my SCIENCE notebook

Your Investigation

Now it's your turn to do your investigation and write about it. Write about the following checklist items in your science notebook.

- ☐ **Choose a question or make up one of your own.**
- ☐ **Gather the materials you will use.**

Model

My Prediction

If the air is humid, then water vapor will condense quickly on the outside of the can. If the air is dry, then it will take longer for water vapor to condense on the can. If the humidity stays the same from day to day, then water vapor will condense at about the same rate.

You can use "If..., then...." statements to make your prediction clear.

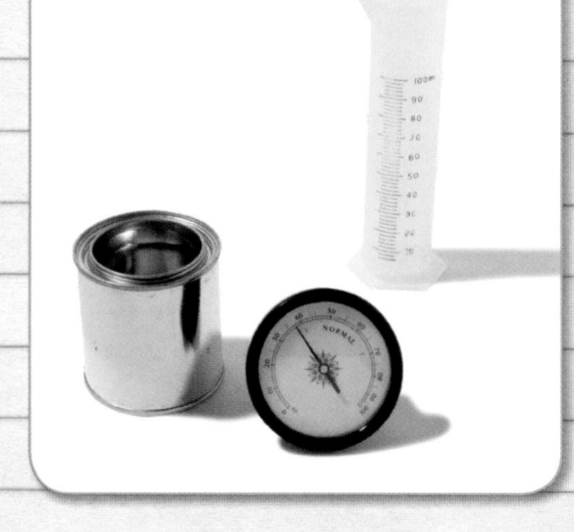

my SCIENCE notebook Your Investigation

☐ **If needed, make a hypothesis or prediction.**

Write your hypothesis or prediction in your science notebook.

Model

Variables I Will Observe or Measure

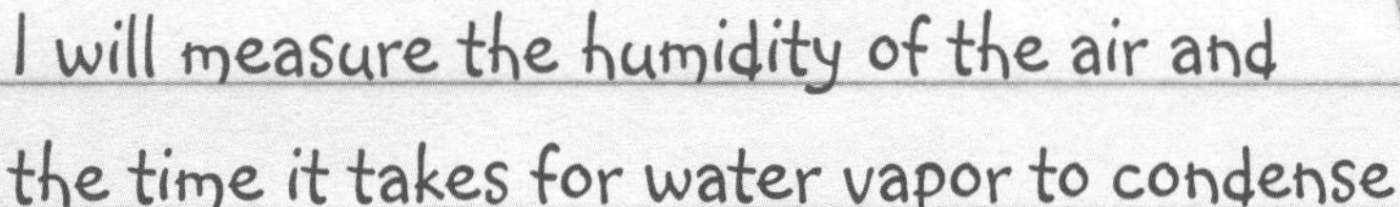

I will measure the humidity of the air and the time it takes for water vapor to condense on the outside of the can.

Variables I Will Keep the Same

Everything else will be the same.
I will use the same amount of water in each trial. I will measure the temperature of the water to make sure it is the same for each trial.
I will use the same can each time.

my SCIENCE notebook Your Investigation

☐ **If needed, identify, manipulate, and control variables.**

Write about the variables for your investigation.

Answer these questions:
What will I observe or measure?
What things will I keep the same?

Model

My Plan

1. Measure the amount of water vapor in the air with the hygrometer.
2. Measure 200 mL of ice water in the graduated cylinder. Pour the ice water into the can.
3. Use the thermometer to measure the temperature of the ice water.
4. Observe the outside of the can. Use the stopwatch to measure how much time passes before water vapor condenses on the outside of the can.
5. Repeat steps 1–4 each day for 4 more days.

Write detailed plans. Another student should be able to repeat your investigation without asking any questions.

Your Investigation

☐ **Make a plan for your investigation.**

Write the steps for your plan.

Model

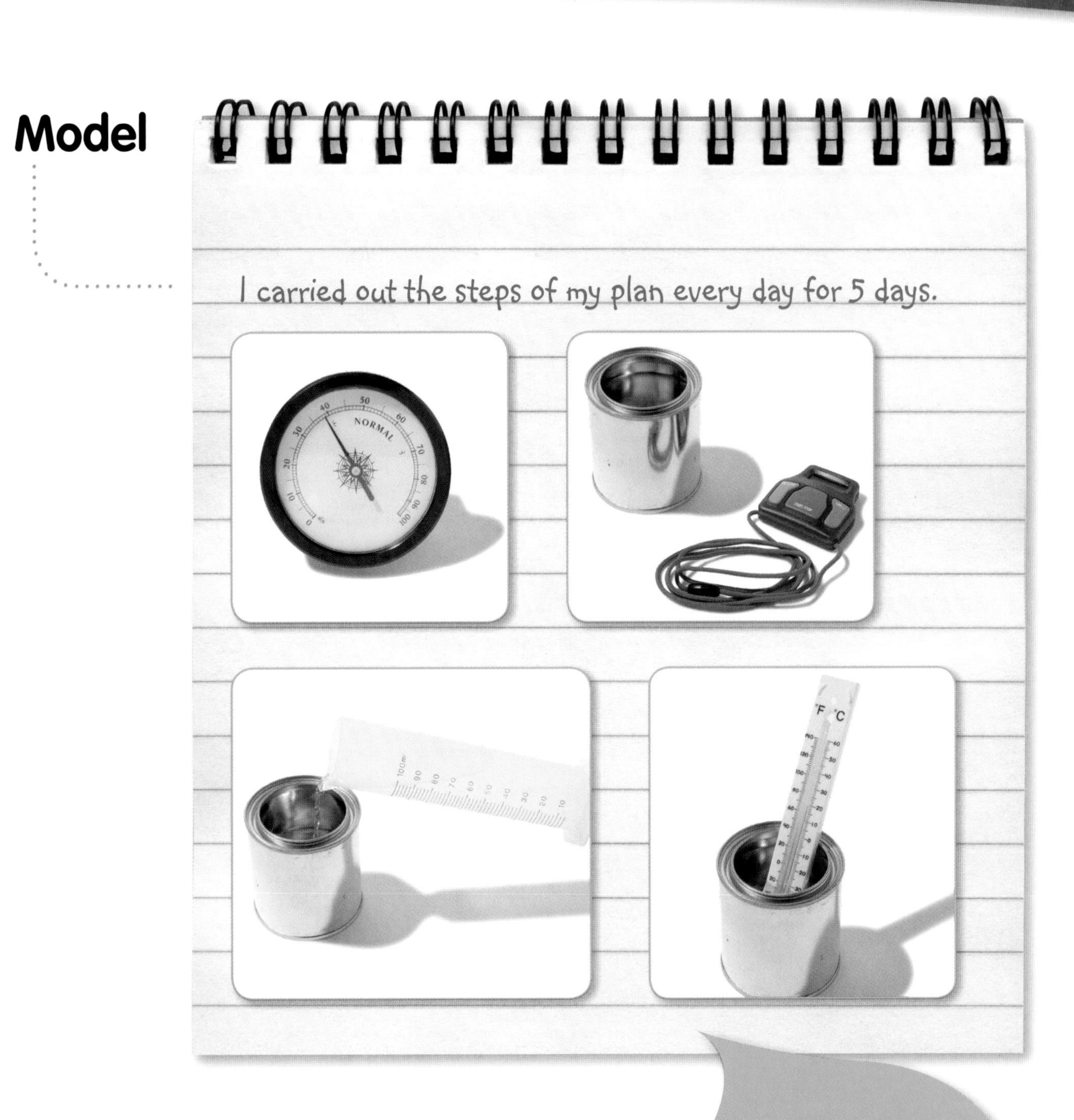

Your Investigation

☐ **Carry out your plan.**

Be sure to follow your plan carefully.

Model

Data (My Observations)

Condensation of Water Vapor on the Can

	Humidity (%)	Temperature of Water (°C)	Time Needed for Condensation to Form
Day 1	58%	1°	15 sec
Day 2	55%	1°	21 sec
Day 3	35%	1°	3 min 25 sec
Day 4	40%	1°	1 min 45 sec
Day 5	31%	1°	3 min 50 sec

My Analysis

The time needed for condensation to form on the outside of the can changed from day to day. Water vapor condensed on the outside of the can more quickly on days with higher humidity than on days with lower humidity.

Explain what happened based on the data you collected.

my SCIENCE notebook

Your Investigation

☐ **Collect and record data. Analyze your data.**

Collect and record your data, and then write your analysis.

Model

How I Shared My Results

I made a graph of my results. Then I explained the steps of my investigation and my results to the class. I used the graph to show patterns in my data and explain what I learned about humidity and condensation.

My Conclusion

The water vapor condensed more quickly on the can on days with high humidity than on days with low humidity, even though the temperature of the water in the can was the same. I conclude that higher humidity causes water vapor to condense more quickly because there is more water vapor in the air.

Another Question

Will the temperature of the air affect how quickly condensation forms on the can?

my SCIENCE notebook

Your Investigation

- ☐ Explain and share your results.
- ☐ Tell what you conclude.
- ☐ Think of another question.

Investigations often lead to new questions for inquiry.

Think Like a Scientist

How Scientists Work

Evaluating Scientific Information

Scientists conduct different types of investigations to answer questions they have about the natural world. But they usually cannot answer their questions based on the results of one investigation. Scientists often do research to gather information that will help them learn more about their research questions. They can review another scientist's observations, or they can analyze the results of other experiments.

This scientist studies whale photos taken by another scientist in the Bay of Fundy, Canada.

When scientists do research, they evaluate information to make sure that it is supported by data. If a claim or explanation is not supported by evidence, scientists will not use it to answer their research questions. Opinions, or beliefs that are not supported by evidence, are not accepted as scientific information.

Scientists must also evaluate whether the data that are being presented are accurate. They analyze the data and judge whether there might have been errors when data were collected or recorded. After they evaluate the strengths and weaknesses of the evidence, they decide whether the data support the scientific explanation that is given.

If data are entered incorrectly into a computer program, the computer analysis of the data might not be correct. Scientists have to evaluate whether the information in a computer program is accurate.

To make sure that information is accurate and reliable, scientists can check multiple sources to see if the data are consistent. Scientists also think about whether there are enough data to fully support a conclusion. Often, they might find that more research needs to be done to make an accurate claim. They may choose to design an investigation that will provide more information.

When scientists do research, they have to keep an open mind about the information they find. Sometimes, the evidence a scientist collects does not answer the question the way he or she expected. When that happens, the scientist might ask a different question or form a new hypothesis.

SUMMARIZE

What Did You Find Out?

1. Why are opinions not accepted as scientific information?

2. Why might scientists check multiple sources to find information?

Practice Evaluating Scientific Information

A team of scientists wanted to find out how the weather in a tropical rainforest affected a particular species of tree frog. The scientists measured the weather patterns in the rainforest for 20 years. They also studied the changes in the tree frog population during those years. Based on the evidence they collected, the team concluded that the weather in the rainforest had become hotter and drier, and that this may be associated with the decline of the tree frog population.

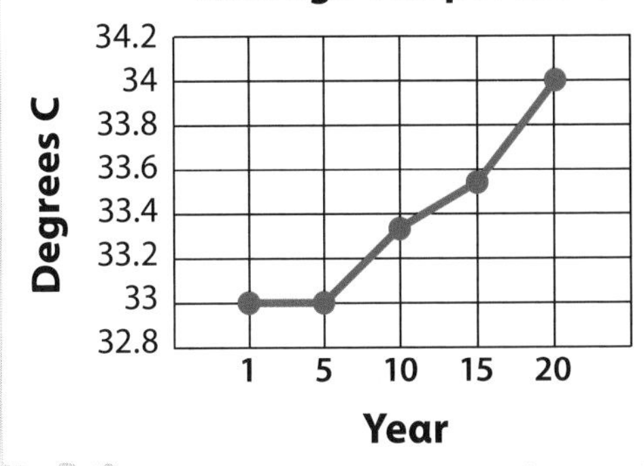

The graphs summarize the data collected by the scientists over the 20 years of their study.

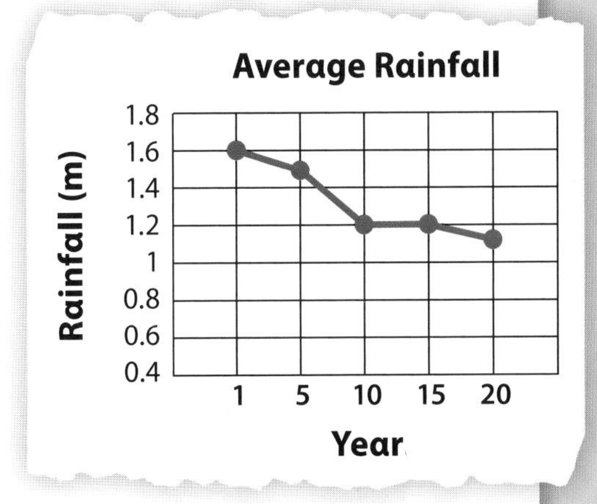

1. Is the information in the graphs fact or opinion? How do you know?

2. Are there other possible causes that would explain the change in the tree frog population? How could you find out?

3. Does the data support the scientists' conclusion? Why or why not?

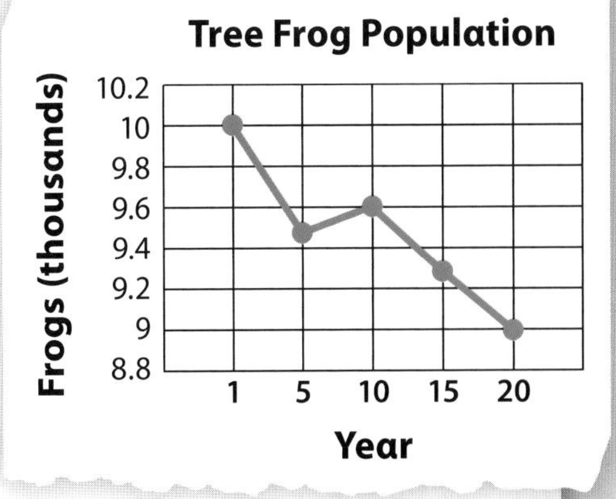

Science in a Snap!

CHAPTER 1

Science in a Snap! Predict What a Magnet Will Attract

Observe a mixture of objects. **Predict** which objects will be attracted to a magnet. Then move a bar magnet over the objects. Separate those that the magnet attracts. How else could you have separated this mixture?

CHAPTER 2

Science in a Snap! Observe a Chemical Change

Dampen a paper towel with vinegar. Place several pennies on the paper towel. Check the pennies the next day. **Observe** how the pennies have changed. What do you think caused the change?

CHAPTER 3

Science in a Snap! Observe Forces on a Penny

Hold a penny above a table and let it go. Then lay the penny flat on a ramp made from a book. Push it down the ramp. Which force was a contact force? Which was a non-contact force?

CHAPTER 4

Science in a Snap! Use a Lever to Raise an Object

Place a whiteboard eraser on a table and balance a ruler on it. Put a wooden block on one end of the ruler. Push down on the other end of the ruler to lift the block. **Investigate** how the position of the eraser affects how hard you have to push down to lift the block. Move the eraser closer to your hand and lift the block. Then lift the block with the eraser farther from your hand. How does the force used to lift the block change? What are other ways you could use a tool to lift the block?

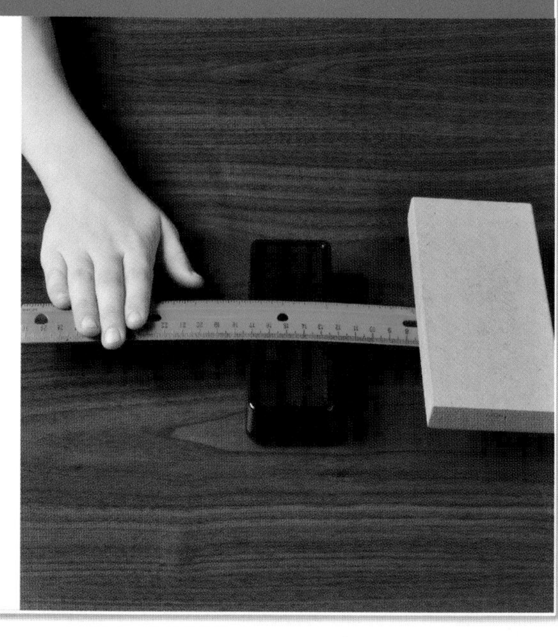

CHAPTER 5

Science in a Snap! Observe Changing Sounds

Hold a ruler flat against a table with half the ruler over the edge. Pluck the end of the ruler and listen to the sound. Press your ear to the table and pluck the ruler again. How does the sound change? **Investigate** how the sound changes if more or less of the ruler is over the edge. Then use more force to pluck the ruler. How does the sound change?

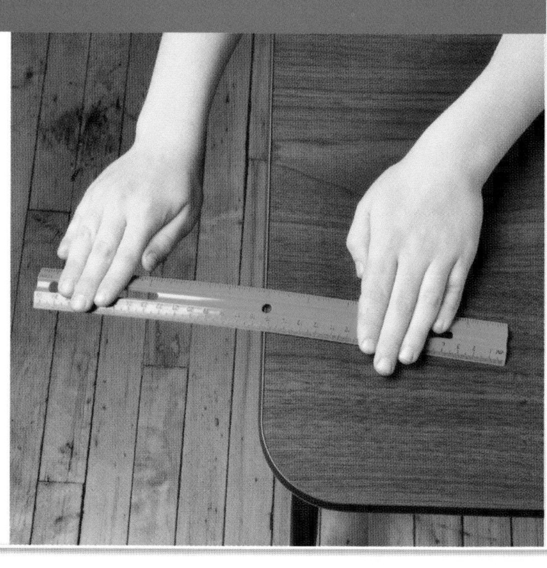

CHAPTER 6

Science in a Snap! Observe an Energy Change

The battery in a flashlight changes chemical energy to electrical energy. How does the electrical energy change? To find out, turn on a flashlight. Hold the bulb of a thermometer against the bulb of the flashlight. **Observe** what happens to the temperature. Electrical energy in the flashlight changes to what two types of energy?

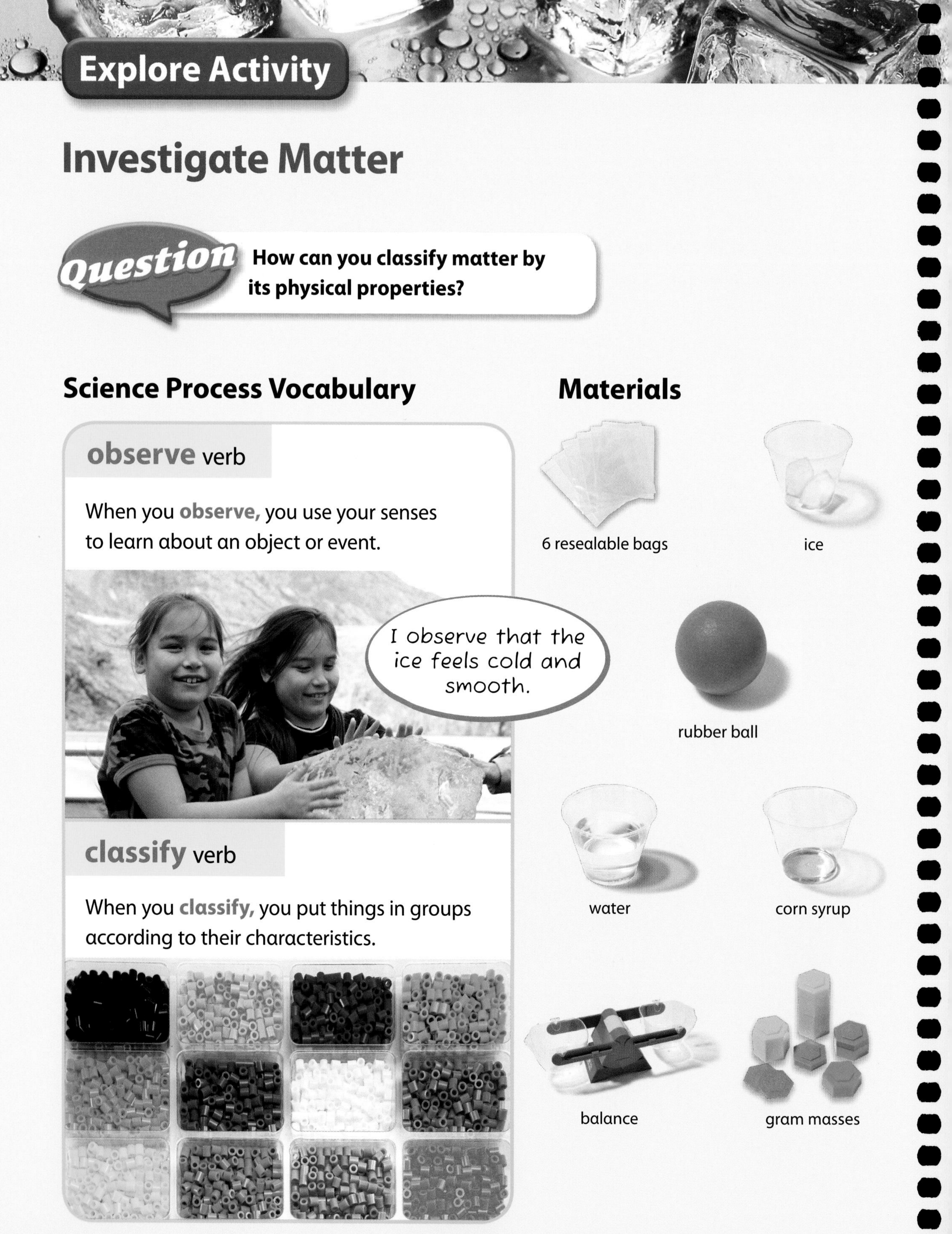

Explore Activity

Investigate Matter

Question **How can you classify matter by its physical properties?**

Science Process Vocabulary

observe verb

When you **observe,** you use your senses to learn about an object or event.

classify verb

When you **classify,** you put things in groups according to their characteristics.

Materials

- 6 resealable bags
- ice
- rubber ball
- water
- corn syrup
- balance
- gram masses

What to Do

1. Put the ice, the rubber ball, the water, and the corn syrup into separate plastic bags. Seal the top of each bag. Fill another plastic bag with air, and seal the top.

2. Use a balance to **measure** the mass of an empty plastic bag. Record the mass of the empty bag in your science notebook.

What to Do, continued

3. Measure and record the mass of the ice and bag. Find the mass of just the ice by subtracting the mass of the empty bag from the mass of the ice and bag. Record the mass of the ice.

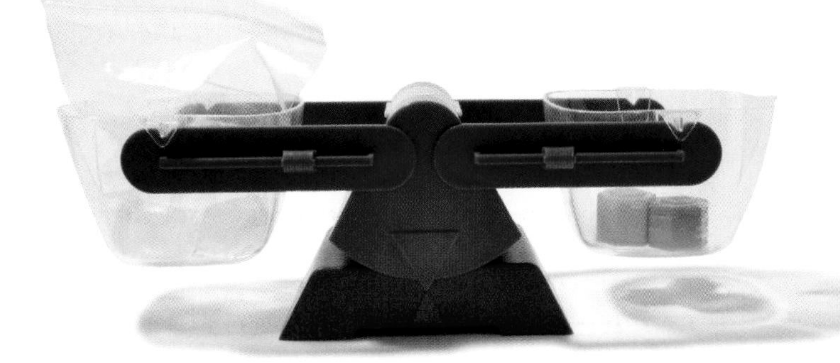

4. Repeat step 3 for each of the other bags of matter.

5. **Observe** other properties of the matter in each bag. Think about the color, texture, and shape of the material. Does the material flow? Does it take the shape of its container? Record your observations.

6. **Compare** the properties of the different materials. **Classify** each material as a solid, a liquid, or a gas.

Record

Write in your science notebook. Use a table like this one.

my SCIENCE notebook

Properties of Matter

Matter in Bag	Mass of Empty Bag (g)	Mass of Matter and Bag (g)	Mass of Matter (g)	Other Properties	Solid, Liquid, or Gas?
Ice					
Rubber ball					
Water					

Explain and Conclude

1. **Compare** the properties of each type of matter you **observed.** How are they alike and different?
2. Which properties helped you **classify** matter as a solid, a liquid, or a gas? Explain.

Is the juice a solid, a liquid, or a gas?

Directed Inquiry

Investigate Compounds

Question **How can you separate hydrogen and oxygen in water?**

Science Process Vocabulary

observe verb

When you **oberve,** you use your senses to tell about an object or event.

compare verb

When you tell how objects or events are alike and different, you **compare.**

The bubbles are different sizes.

Materials

safety goggles

baking soda

spoon

cup with water

2 insulated pieces of wire with ends stripped

9-volt battery

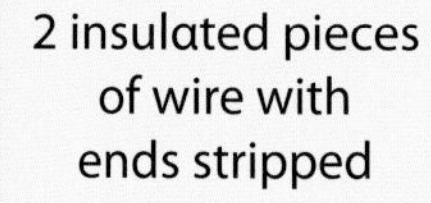

2 pencils, sharpened at both ends

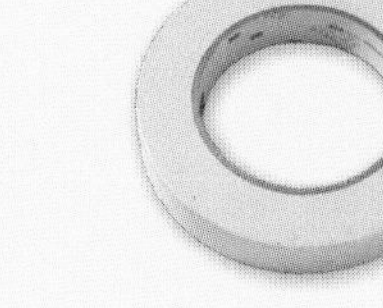

tape

index card with holes

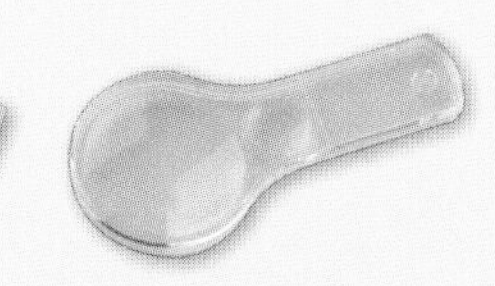

hand lens

What to Do

1. Put on your safety goggles. Add a spoonful of baking soda to the cup of water. Stir until the baking soda dissolves.

2. Connect one end of a wire to the terminal of the battery. Connect the other wire to the other terminal. Tape the wires firmly in place.

3. Wrap the other end of a wire around the graphite point of a pencil. Be sure the stripped end of the wire touches the graphite. Wrap the end of the other wire around the end of the second pencil. Tape the wires firmly in place.

What to Do, continued

4. Push the free ends of the pencil through the holes in the index card. Place the card on the cup so that the pencil points are in the water. Electricity is now flowing through the water.

5. When electricity flows through water, it separates the hydrogen and oxygen in the water. Use a hand lens to **observe** the bubbles of hydrogen and oxygen gas that are formed on the ends of the pencils. Each bubble contains many atoms. **Compare** the bubbles that form on the pencil points. Record your observations in your science notebook. Then disconnect the wires from the battery and the pencils.

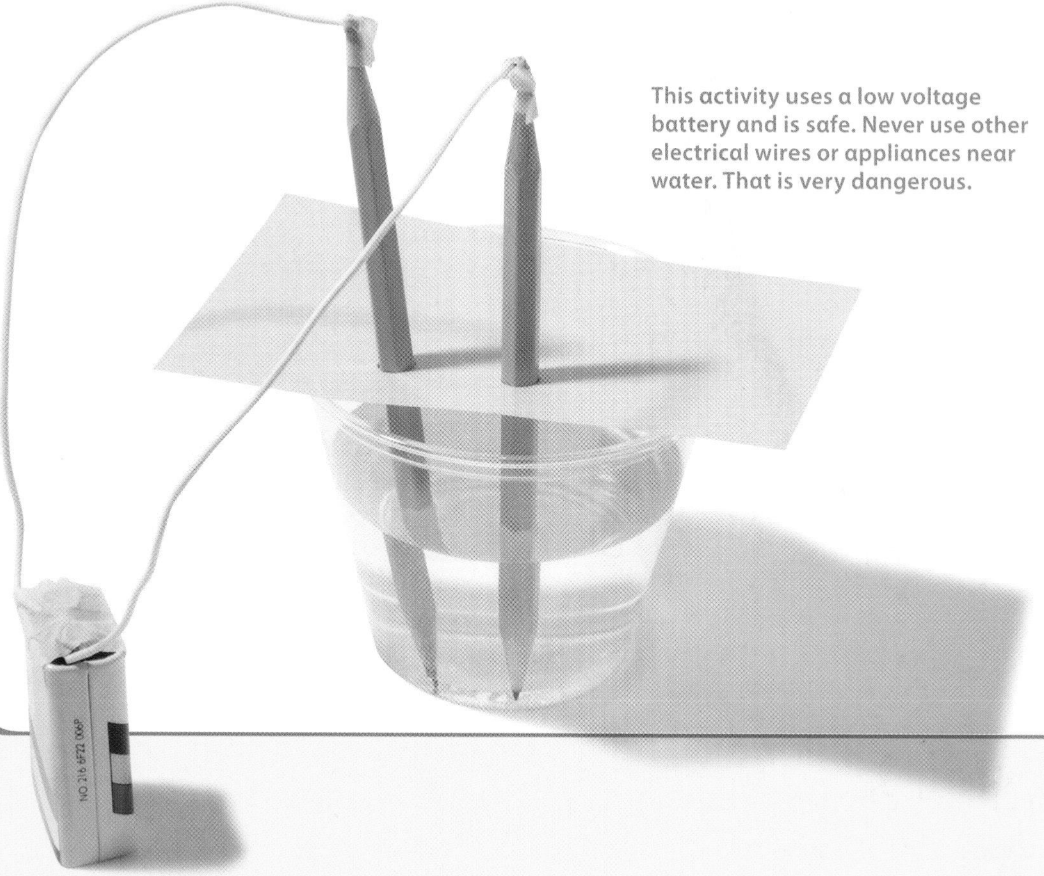

This activity uses a low voltage battery and is safe. Never use other electrical wires or appliances near water. That is very dangerous.

Record

Draw your set-up in your science notebook. Show the bubbles forming at each pencil in the water.

my SCIENCE notebook

Bubbles at Pencil Tips

Explain and Conclude

1. **Compare** the amounts of bubbles produced in the water at each pencil.
2. Where did the hydrogen and oxygen gas that formed the bubbles come from?
3. In water, there are 2 atoms of hydrogen for every 1 atom of oxygen. **Infer** at which pencil hydrogen gas is forming. What evidence did you use to make your inference?

Think of Another Question

What else would you like to find out about separating hydrogen and oxygen in water? How could you find an answer to this new question?

Directed Inquiry

Investigate Physical Changes in Water

Question **How does temperature affect the physical properties of water?**

Science Process Vocabulary

measure verb

When you **measure,** you find out how much or how many.

data noun

Data are observations and information that you collect and record in an investigation.

I can use a thermometer to collect data about the temperature of water.

Materials

jar with lid

balance

gram masses

water

graduated cylinder

What to Do

1. Place the empty jar and lid on one balance pan. Add gram masses to the other pan until the pans are balanced. Record the mass of the jar and lid in your science notebook.

2. Use the graduated cylinder to **measure** 50 mL of water. Pour the water into the jar. **Observe** the physical properties of the water. Record your observations. Put the lid on the jar.

What to Do, continued

3. Use the balance to find the mass of the jar, lid, and water. To find the mass of the water, subtract the mass of the jar and the lid from the mass of the jar, lid, and water. Record your **data.**

4. Place the jar with water in a freezer. **Predict** what will happen to the mass of the water when it freezes.

5. After all the water has frozen, use the balance to find the mass of the jar, lid, and ice. To find the mass of the ice, subtract the mass of the jar and the lid from the mass of the jar, lid, and ice. Record your data.

6. Remove the lid from the jar. Observe the physical properties of the ice. Record your observations.

Record

Write in your science notebook. Use a table like this one.

my SCIENCE notebook

Physical Properties of Water and Ice

	Mass (g)	Properties
Jar and lid		
Jar, lid, and water		
Water		
Jar, lid, and ice		

Explain and Conclude

1. Did your results support your **prediction?** Explain.
2. **Compare** the mass and other physical properties of the ice and the water. Which properties changed and which did not?

Think of Another Question

What else would you like to find out about how temperature affects the physical properties of water? How could you find an answer to this new question?

Jokulsarlon, Iceland

Guided Inquiry

Investigate a Chemical Reaction

Question How can you change how fast a chemical reaction happens?

Science Process Vocabulary

hypothesis noun

You make a **hypothesis** when you state a possible answer to a question that can be tested by an experiment.

If I place a tablet in water, then the water and tablet will react.

variable noun

A **variable** is part of an experiment that you can change.

I will only change one thing to see if it changes how fast the reaction happens.

Materials

safety goggles

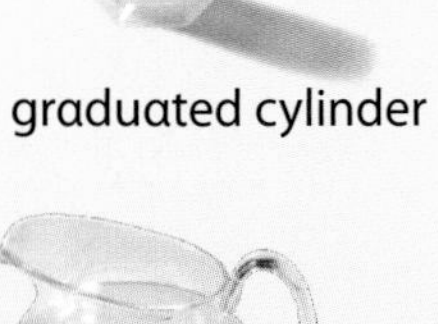

graduated cylinder

room-temperature water

cold water

warm water

4 plastic cups

thermometer

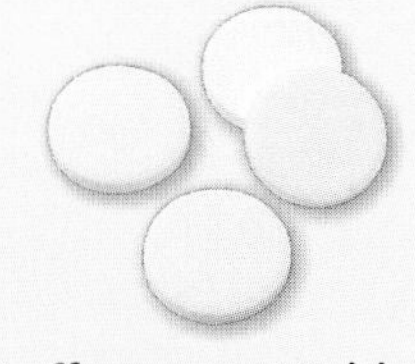

4 effervescent tablets

stopwatch

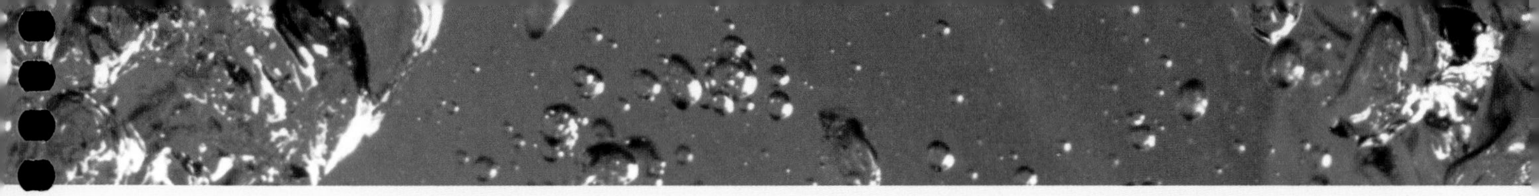

Do an Experiment

Write your plan in your science notebook.

Make a Hypothesis

In this investigation, you will place an effervescent tablet in water and measure the time it takes for a complete reaction. Then you will either warm or cool the water. How will this temperature change affect the rate of reaction? Write your **hypothesis.**

Identify, Manipulate, and Control Variables

Which variable will you change?
Which variable will you observe or measure?
Which variables will you keep the same?

What to Do

1. Put on your safety goggles. Use the graduated cylinder to **measure** 100 mL of room-temperature water. Pour the water into a plastic cup. Use a thermometer to measure the temperature of the water. Record the **data** in your science notebook.

2. **Predict** how long it will take one tablet to completely react with the water. Record your prediction.

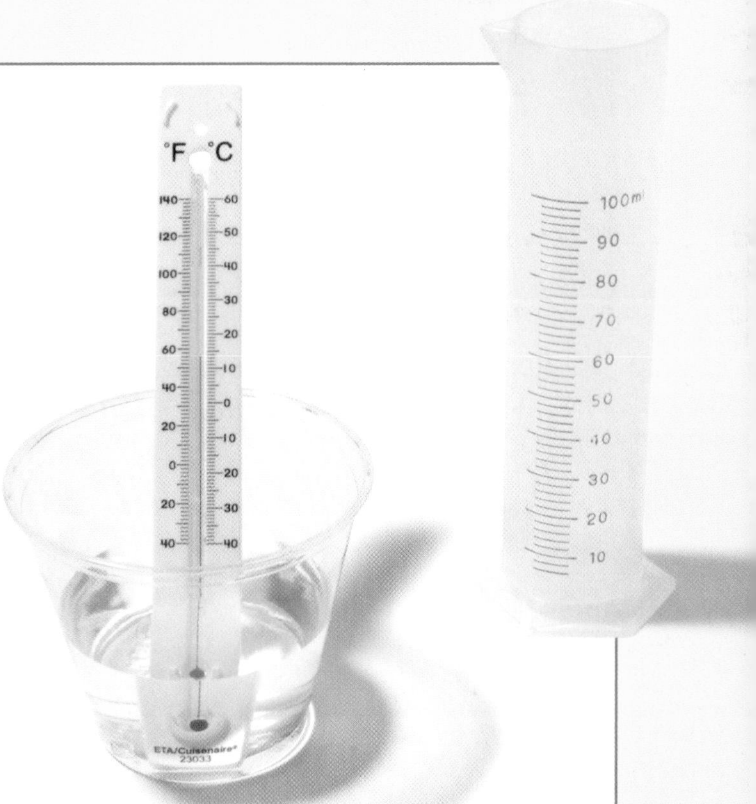

What to Do, continued

3. Put the tablet in the water and **observe** the reaction. Use a stopwatch to measure the time it takes for the tablet to stop producing bubbles. Record your data.

4. Repeat steps 1–3, this time using either warm or cold water. Record your data.

5. Repeat step 3 two more times using the same water temperature you used in step 4.

Record

Write in your science notebook.
Use a table like this one.

my SCIENCE notebook

Temperature and Reaction Time

Water Temperature (°C)	Predicted Time (s)	Actual Time (s)
Room temperature ______		
Trial 1 ______		
Trial 2 ______		
Trial 3 ______		

Explain and Conclude

1. Do the results support your **hypothesis?** Explain.
2. Did you get the same results for all 3 trials? If not, what might have caused the difference?
3. **Share** your results with those of other groups. What can you **conclude** about temperature and how fast a chemical reaction happens?

Think of Another Question

What else would you like to find out about how temperature affects how fast a chemical reaction happens? How could you find an answer to this new question?

Keeping food cold slows down spoiling, which is a chemical reaction.

Think
Like a Scientist

Math in Science

Measuring Forces

When you describe a force, you tell about both the size of the force and its direction. For example, you might say, "The person in the picture is pulling the wagon to the left with a strong force."

However, in science, using exact measurements is important. You can use a spring scale to measure the size of a force. The unit used to measure force is the newton (N).

Adjustment nut

The picture shows a spring scale. Notice the spring inside. When a force pulls on the hook of the spring scale, the force causes the spring inside to stretch. The amount the spring stretches depends on how strong the force is. A strong force will cause the spring to stretch more than a weak force.

A scale is used to measure how many newtons of force are pulling on the spring. Different spring scales can have different scales. Some spring scales measure up to 5 N, while others can measure up to 1,000 N.

Spring

Indicator disk

Scale

Notice that the red scale on the left in the picture goes from 0 N to 20 N. Each line between two numbers stands for 0.5 N. The green scale on the right goes from 0 N to 5 N. Each line between its numbers stands for 0.1 N.

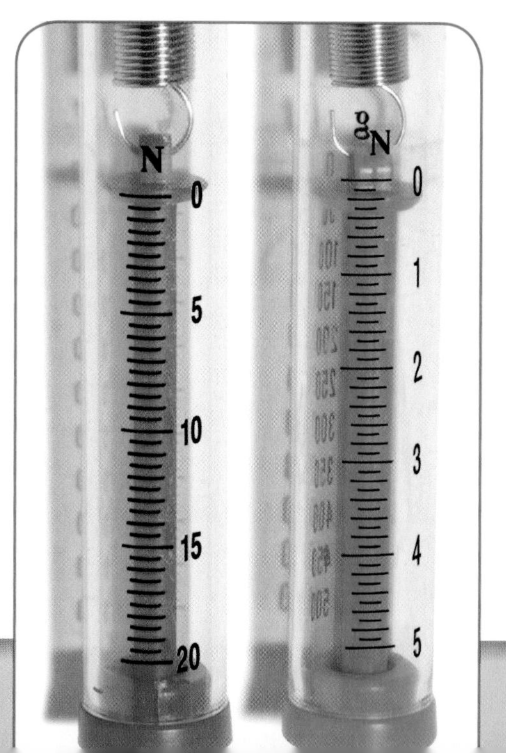

Using a Spring Scale

To measure the force applied to an object, follow these steps.

1. Make sure the top of the indicator disk is at the 0 mark on the scale.
If not, turn the adjustment nut on the top of the spring scale until the disk is at 0.

2. Attach the object to the hook of the spring scale.

3. Pull on the opposite end of the spring scale to move the object.

4. Read the number of newtons on the spring scale.

SUMMARIZE

What Did You Find Out?

1. What is a newton?

2. How does a spring scale work?

Practice Reading a Spring Scale

Attach a spring scale to a wooden block. Pull the block 5 times with different amounts of force. Have a partner tell how many newtons of force you apply each time.

Then have the partner pull the block 5 times with different forces as you read the spring scale. How does the number of newtons change as you or your partner pull harder?

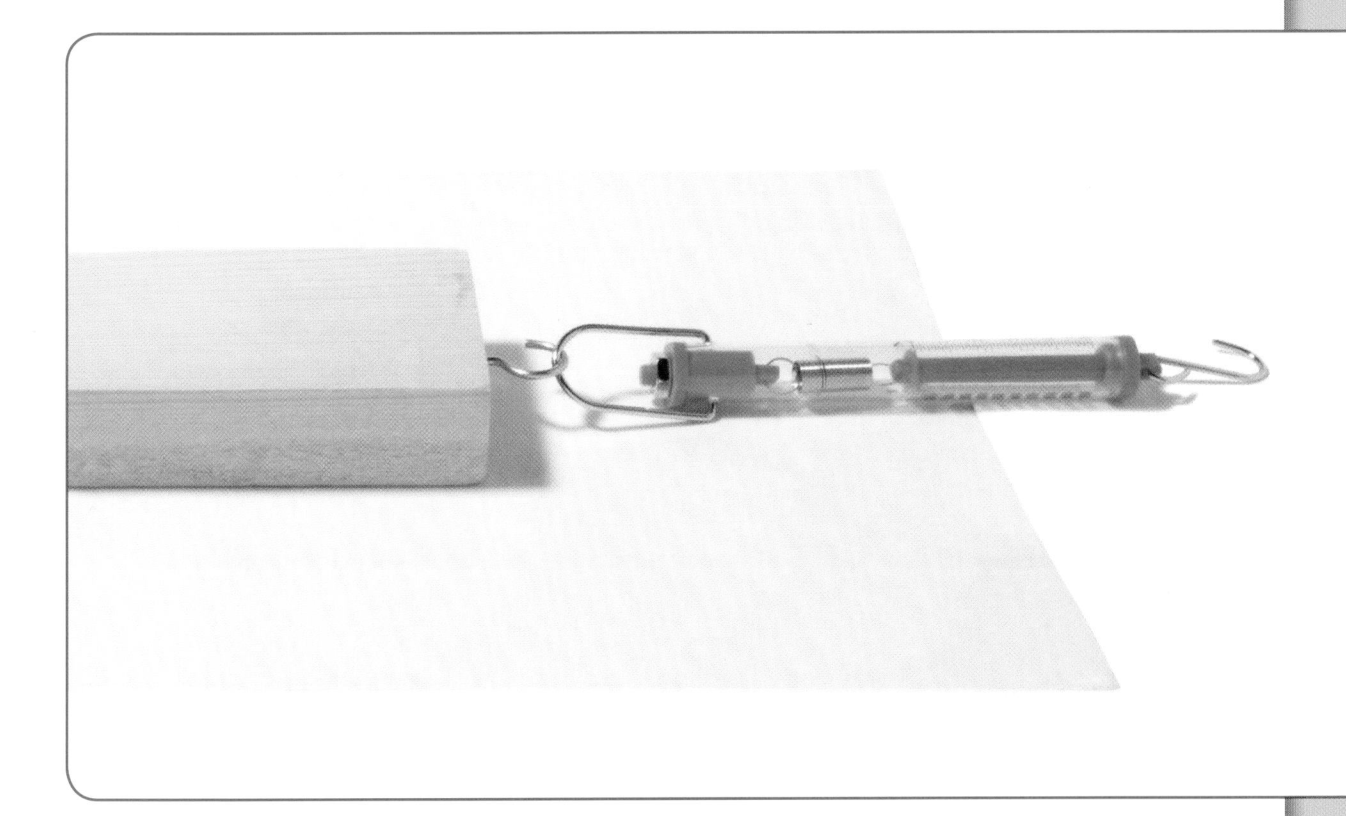

Directed Inquiry

Investigate Forces and Motion

Question **What happens to the motion of a wooden block when forces act on it?**

Science Process Vocabulary

measure verb

When you **measure,** you can use tools to find out how much or how many.

conclude verb

You **conclude** when you use information, or data, from an investigation to come up with a decision or answer.

I measured force on the spring scale every time I moved the block. I conclude that force is needed to move the block.

Materials

safety goggles

waxed paper

tape

wood block with a hook at each end

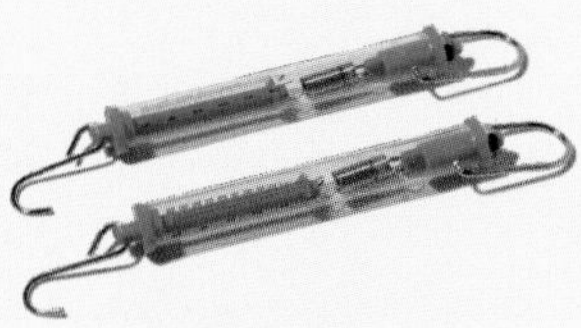
2 spring scales

What to Do

1. Put on your safety goggles. Tape a sheet of waxed paper to a table. Place the wood block on the waxed paper. Attach a 5-N spring scale to each end of the block.

2. How can you and your partner both pull on the spring scales so that the block does not move? How will the forces you and your partner use **compare?** Write your **prediction** in your science notebook.

3. Pull each spring scale in opposite directions away from the block with 1 N of force. What happens to the block? Record the forces and your **observations.**

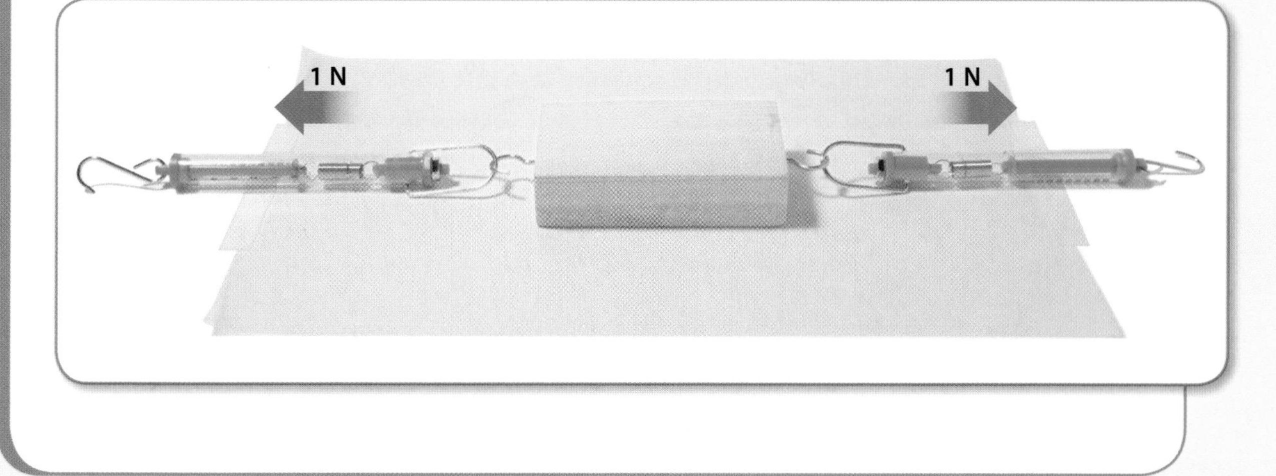

What to Do, continued

4. Then pull each spring scale away from the block with 2 N of force. Record your observations.

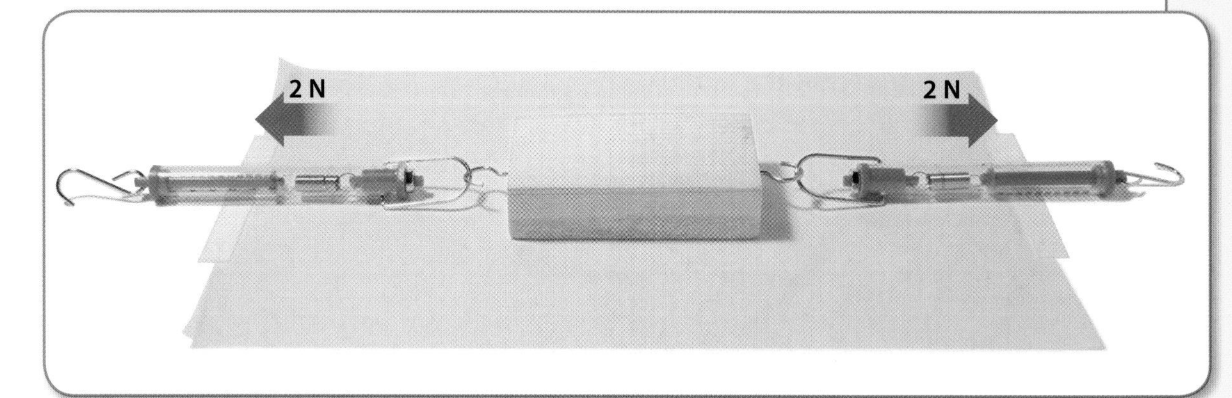

5. Pull on the spring scales again. This time the partner on the right should use more force. **Measure** the forces you and your partner use. Record the **data.** Also record which way the block moves.

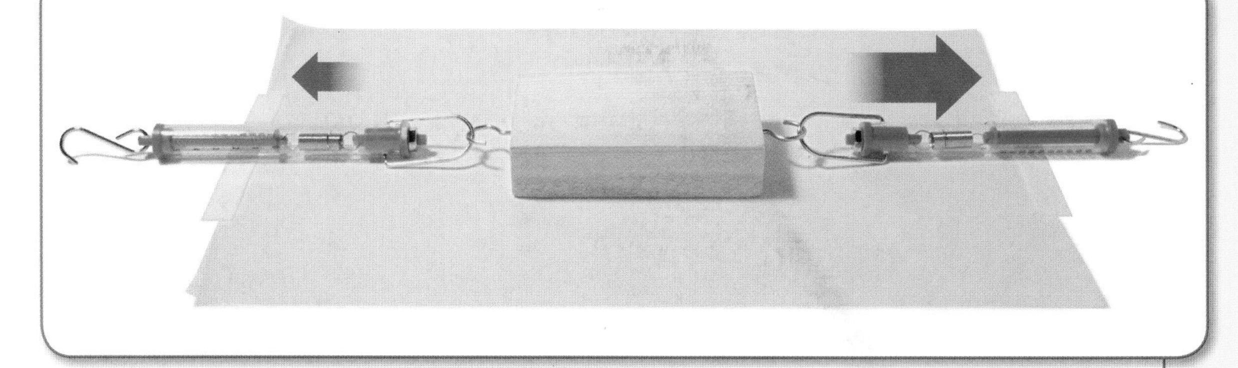

6. Repeat step 5, but this time the partner on the left should pull harder.

Record

Write in your science notebook. Use a table like this one.

my SCIENCE notebook

Forces and Motion

Force on Left Spring Scale	Force on Right Spring Scale	Direction Block Moves
1 N	1 N	
2 N	2 N	

Explain and Conclude

1. Did your results support your **prediction?** Explain.
2. **Compare** the forces that caused the block to move in step 5 and in step 6. What can you **conclude** from this **data** about forces and the motion of the block?
3. How are the forces in steps 5 and 6 different from those in steps 3 and 4? How did the difference affect the motion of the block?

Think of Another Question

What else would you like to find out about how forces and motion are related? How could you find an answer to this new question?

Forces acting on this umbrella cause it to move.

Guided Inquiry

Investigate Changes in Motion

Question How do changes in force and mass affect an object's motion?

Science Process Vocabulary

predict verb

When you **predict,** you use observations and what you already know to tell what you think will happen.

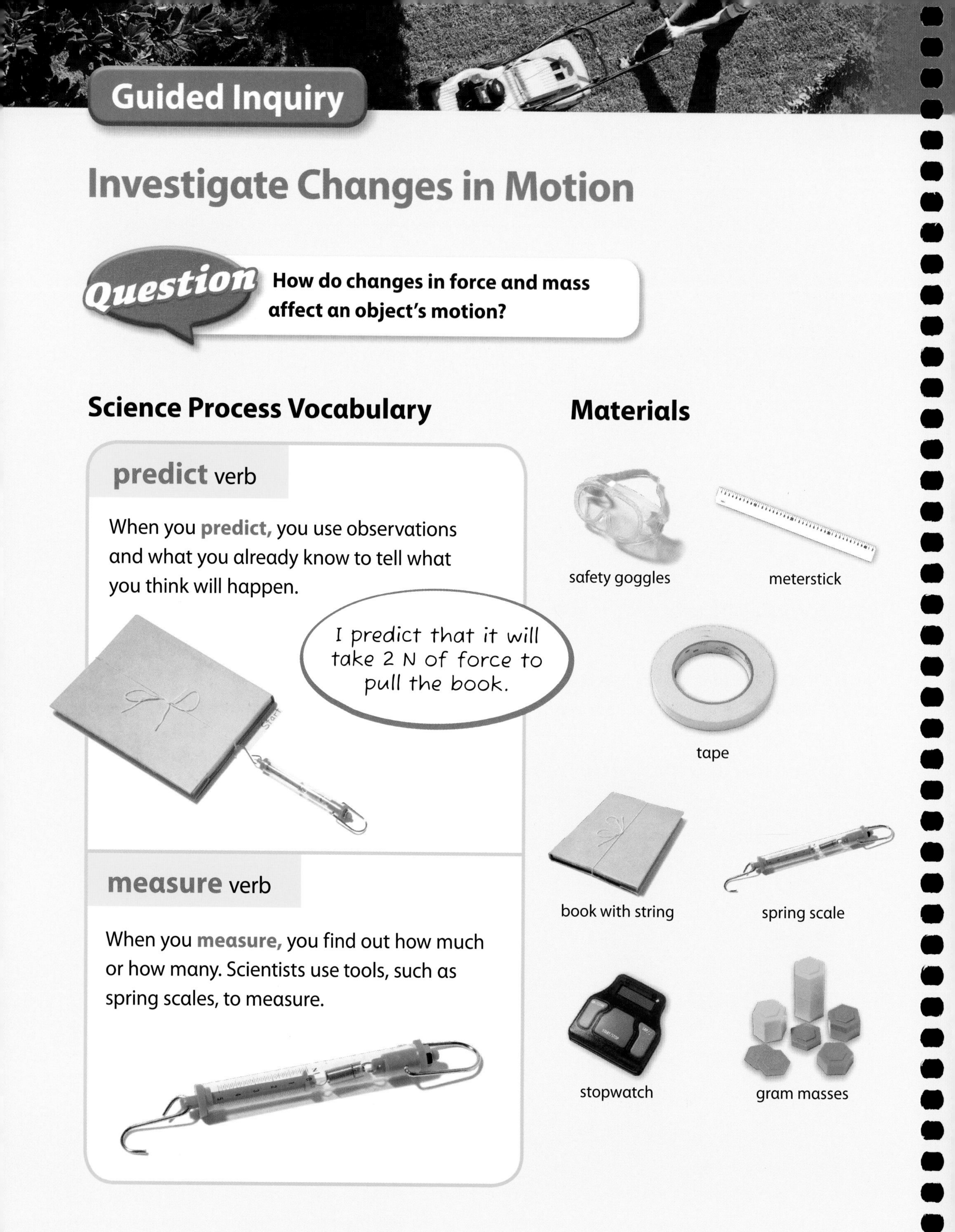

measure verb

When you **measure,** you find out how much or how many. Scientists use tools, such as spring scales, to measure.

Materials

safety goggles

meterstick

tape

book with string

spring scale

stopwatch

gram masses

Do an Experiment

Write your plan in your science notebook.

Make a Hypothesis

In this investigation, you will measure the time required to pull a book a certain distance. You can change the mass of the book by adding gram masses. Or, you can change the amount of force you use when pulling the book. How will the change in mass or force affect the motion of the book? Write your **hypothesis.**

Identify, Manipulate, and Control Variables

Which variable will you change?
Which variable will you observe or measure?
Which variables will you keep the same?

What to Do

1. Put on your safety goggles. Use a meterstick to **measure** 2 points on a table that are 1 m apart. Use tape to mark each point. Label the pieces of tape **Start** and **Stop.**

Start
1 meter
Stop

2. Place the book with string behind the piece of tape labeled **Start.** Hook the spring scale to the string.

What to Do, continued

3 **Predict** how much time it will take to pull the book from **Start** to **Stop** using 3 N of force. Use a stopwatch to measure the time it takes to pull the book to the **Stop** point. Record the **data** in your science notebook.

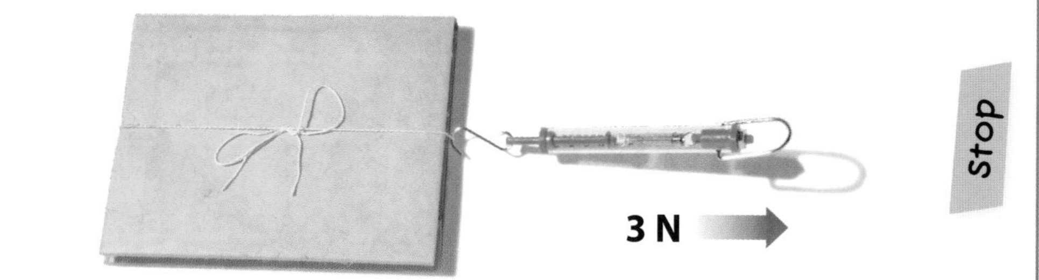

4 Decide with your group whether you will change the mass you will pull or the amount of force you will use.

5 If you choose to change the mass, decide how much mass you will add on top of the book. Record the mass you added. Predict how much time will be needed to pull the book from **Start** to **Stop** using 3 N of force. Then measure and record the actual time it takes to pull the book.

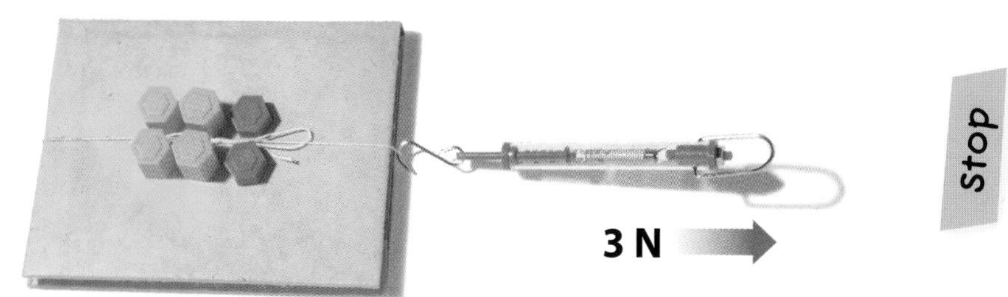

6 If you choose to change the force, decide on a different amount of force to pull the book. Predict how much time will be needed to pull the book with the different force. Pull the book from **Start** to **Stop.** Measure and record the actual time it takes to pull the book.

Record

Write in your science notebook.
Use a table like this one.

my SCIENCE notebook

Force, Mass, and Motion

What I Pulled	What I Changed	Predicted Time (s)	Measured Time (s)
Book			

Explain and Conclude

1. Do the results support your **hypothesis?** Explain.
2. What **variable** did you test in this experiment?
3. **Share** your results with your classmates. **Conclude** how the motion of an object is affected by changing the object's mass or the force used to move it.

Think of Another Question

What else would you like to find out about how changes in force and mass affect an object's motion? How could you find an answer to this new question?

The amount of force needed to move the soil depends on the mass of the soil.

Directed Inquiry

Investigate Friction

Question How does friction affect the force needed to do work over different surfaces?

Science Process Vocabulary

compare verb

When you **compare,** you tell how objects or events are alike and different.

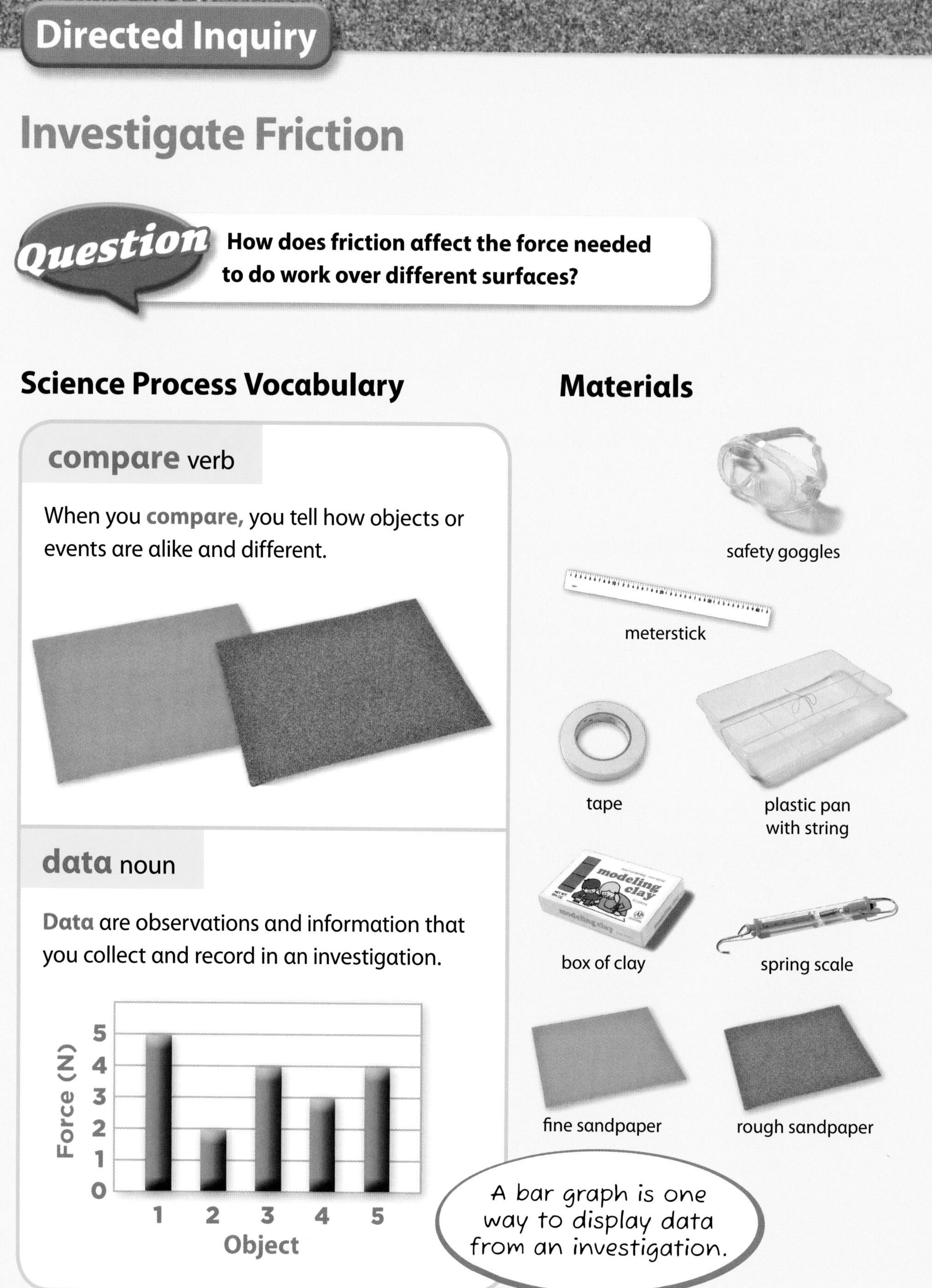

data noun

Data are observations and information that you collect and record in an investigation.

Materials

safety goggles

meterstick

tape

plastic pan with string

box of clay

spring scale

fine sandpaper

rough sandpaper

What to Do

1. Put on your safety goggles. Use a meterstick to **measure** 2 points on a table. The 2 points should be 1 m apart. Use tape to mark the points, and label them **Start** and **Stop.**

2. Put the box of clay in the pan, and place the pan on the **Start** line. Attach the spring scale to the string. Pull the pan with a constant force so that the spring scale reading remains constant. **Observe** the force as you pull. Round the force to the nearest newton. Record the **data** in your science notebook.

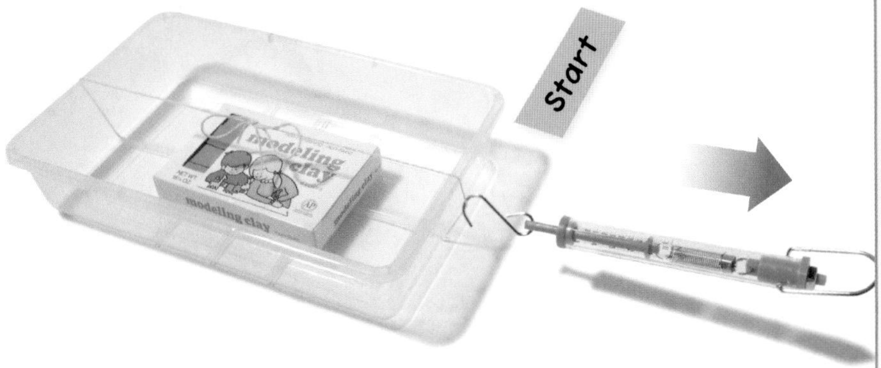

3. Tape a sheet of fine sandpaper to the bottom of the pan. The rough part of the paper should be facing away from the pan. Use rolled pieces of the tape between the sandpaper and the pan so that no tape covers the rough side of the sandpaper.

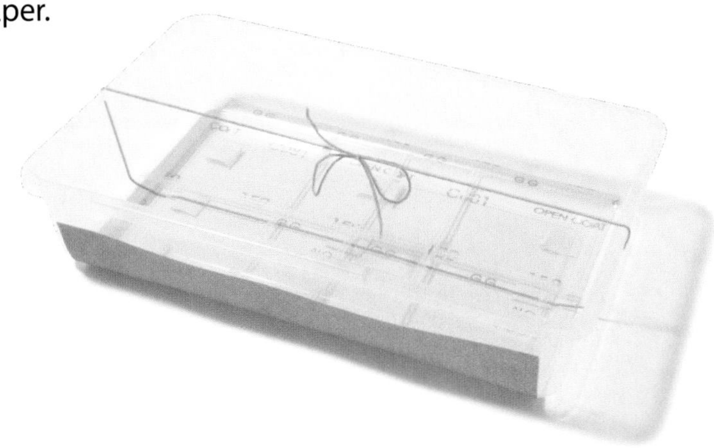

What to Do, continued

4. Use the spring scale to pull the pan with the sandpaper from **Start** to **Stop.** Pull the pan with a steady force so that the spring scale reading remains constant. Measure and record the force you need to move the pan.

5. Replace the fine sandpaper on the pan with rough sandpaper. Use the spring scale to pull the pan from **Start** to **Stop** with a steady force. Record the force needed.

6. Organize your data in a bar graph.

Record

Write in your science notebook. Use a table like this one.

Force Used to Move Pan Across Different Surfaces

Surface	Force (N)
No sandpaper	
Fine sandpaper	
Rough sandpaper	

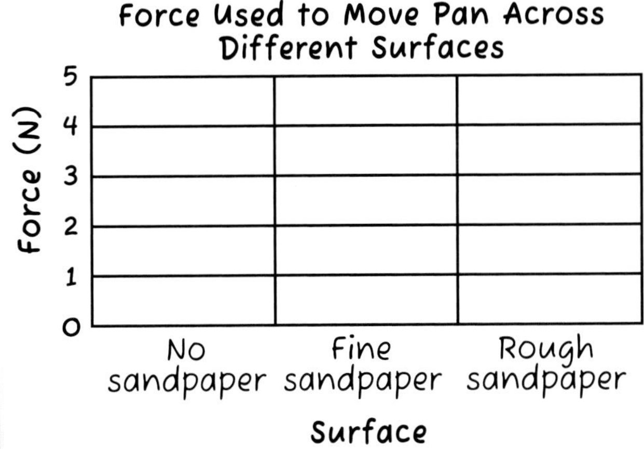

Explain and Conclude

1. **Compare** the force needed to pull the pan on each surface. Which surface required the least force? Which required the most force?
2. Which surface caused more friction?
3. What can you **conclude** about how friction affects the force needed to do work over different surfaces?

Think of Another Question

What else would you like to find out about how friction affects the force needed to do work over different surfaces? How could you find an answer to this new question?

The smooth surface of the snow makes moving the sled easier.

Investigate a Simple Machine

Question **How does the length of a ramp affect the amount of force used to lift an object?**

Science Process Vocabulary

investigate verb

When you **investigate,** you make a plan and follow the plan to answer a question.

data noun

The information you gather during an investigation is called **data.**

I will collect data about forces in my investigation.

Materials

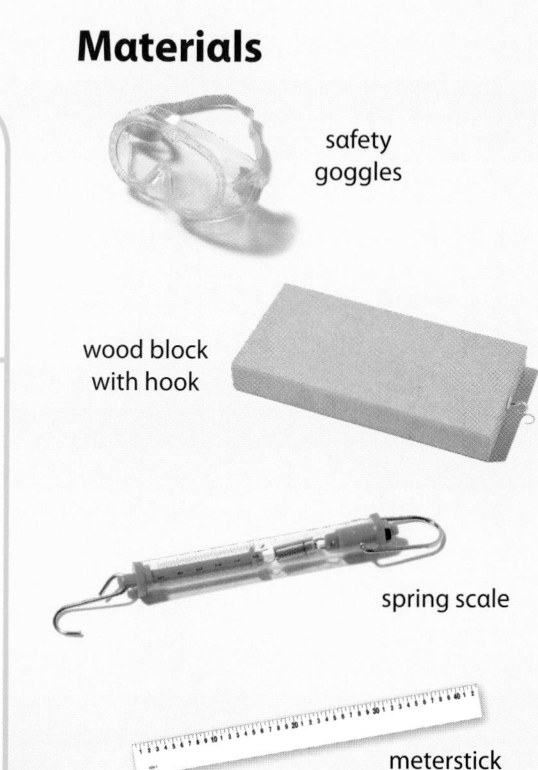

3 pieces of cardboard

Do an Experiment

Write your plan in your science notebook.

Make a Hypothesis

In this investigation, you will design and build two ramps of different lengths to lift a block to a height of 10 cm. How will the length of the ramp effect the force needed to lift the block? Write your **hypothesis.**

Identify, Manipulate, and Control Variables

Which variable will you change?
Which variable will you observe or measure?
Which variables will you keep the same?

What to Do

1. Put on your safety goggles. Attach the spring scale to the block. Hold the block near a meterstick. **Observe** the force as you lift the block 10 cm. Lift with a constant speed. Record the force in your science notebook.

2. Repeat the test 2 more times.

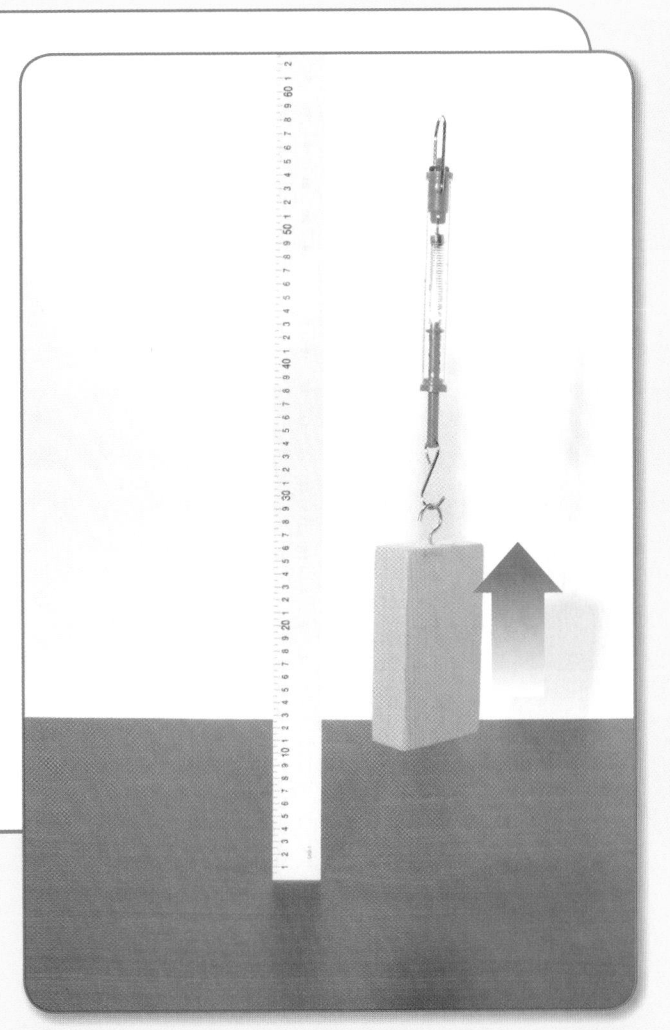

What to Do, continued

3. Next you will pull the wood block up a ramp. Choose a piece of cardboard. Measure and record the length. Then use the cardboard and the books to build a ramp. The ramp should be about 10 cm high.

4. Place the wooden block at the bottom of the ramp. Attach the spring scale to the block. Pull the block up the ramp. Pull at a constant speed that is similar to what you used in step 1. Observe how much force you used to pull the block. Record your **data.**

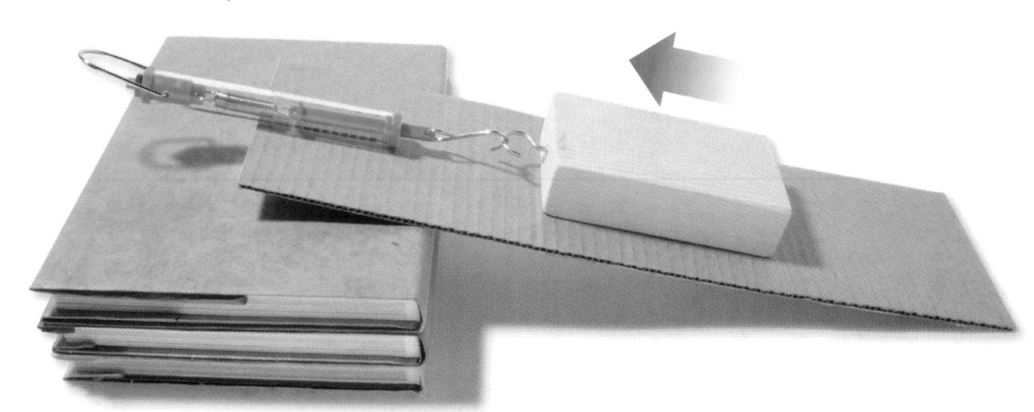

5. Repeat step 4 two more times.

6. Repeat steps 3–5 with a cardboard ramp of a different length.

Record

Write in your science notebook. Use a table like this one.

my SCIENCE notebook

Force Needed to Lift the Block

Setup	Force (N) Trial 1	Force (N) Trial 2	Force (N) Trial 3
Straight up			
Ramp 1 ______ cm			

Explain and Conclude

1. Did your results support your **hypothesis?** Explain.
2. **Compare** your results with the results of other groups. How did the length of the ramp affect the amount of force needed to lift the block?

Think of Another Question

What else would you like to find out about ramps and forces used to lift an object? How could you find an answer to this new question?

This pulley is used to lift a large load.

Think
Like a Scientist

How Scientists Work

Designing Machines to Solve Problems

Scientists and engineers often design and build machines that solve a problem. Every machine does work, and usually this work makes tasks easier for humans. Sometimes, though, machines are designed to help people have fun! For instance, a machine can solve the problem of how to fit a 30-meter water slide in an area 6 meters wide and 9 meters long.

The spiral water slide is a ramp.

Most of the machines you use are compound machines made up of several simple machines. The simple machines work together so that the compound machine can do a job with the least amount of work.

A pizza cutter is a compound machine that solves the problem, "How can you slice a pizza with a diameter that is greater than the length of most knives?"

The pizza cutter is made up of a wedge and a wheel-and-axle.

Robotic arms are compound machines that include wedges, levers, wheels-and-axles, pulleys, and screws. These machines are widely used in manufacturing. They complete repetitive or unsafe tasks.

Once scientists saw how useful robotic arms were, they turned to a new problem: developing a robot that looks and moves like a human. Early robots called Robot Assistants could sweep the floor, collect dirty dishes, load the dishwasher, and move chairs. Scientists tested the robots by giving them simple tasks to perform.

Now scientists are working to adapt the Robot Assistant to be a caregiver for people who need extra help. Scientists want the Robot Assistant to keep track of a patient's vital signs, make sure a patient takes medicines properly, and alert medical staff in case of an emergency.

A robotic arm

Scientists are developing walking, talking robots that look like humans. These robots could act as caregivers for people who need help.

Scientists and engineers use a number of steps when they develop a machine to solve a problem. Many times, they break up the problem into simpler problems. Then they generate solutions for each of the simpler problems.

For example, one step in designing an entire robot was to develop a machine that worked like a forearm. In another step, scientists developed a lever system that functions like our thumb and index finger. Scientists may build and test hundreds of smaller machines. They test each machine using suitable instruments and techniques. They record measurements and other data. Then they assess results and the effectiveness of the machine. If necessary, they modify the machine.

Eventually, they combine the smaller machines to form groups of machines until they have a single complete solution—a freestanding robot.

SUMMARIZE

What Did You Find Out?

1. Why do scientists design machines?

2. What is one step scientists might take when they start to design a machine to solve a problem?

Design a Machine to Solve a Problem

You can use a variety of materials to make a compound machine that solves a problem.

- Choose a problem to solve.
- Design a compound machine that solves the problem. Tell what the machine must do. Describe anything the machine must *not* do.
- Draw a diagram of your design. Label the simple machines that make up your compound machine.
- Record and collect the materials and tools you need to build your machine.
- Build and test your machine.
- Discuss your results with others.
- Change or rework your design to make your machine work better.
- Test again. Record your results.

Name of Machine ____________

Problem to solve	
What the machine must do	
What the machine must not do	
Materials used	
First test results	
Changes needed	
Second test results	

Directed Inquiry

Investigate Mechanical Energy

How does a marble's mechanical energy change as its release height on a ramp changes?

Science Process Vocabulary

measure verb

When you **measure,** you find out how much or how many.

analyze verb

When you **analyze** data, you study it carefully so that you can explain and interpret what has happened.

I can analyze my results to explain how the marble's energy changes.

Materials

safety goggles

3 books

foam tube

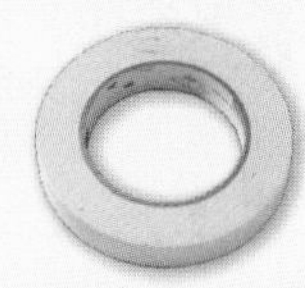

tape

plastic cup

marble

meterstick

What to Do

1. Put on your safety goggles. Place 3 books on the floor. Tape one end of the foam tube to the books. Tape the lower end of the tube to the floor. Place the plastic cup just at the lower end of the tube. The cup should touch, but not overlap, the end of the tube.

2. Place a marble at the top of the tube. Let go of the marble so that it rolls down the tube. Do not push it. **Observe** the marble as it rolls down the tube and into the cup. Record your observations in your science notebook.

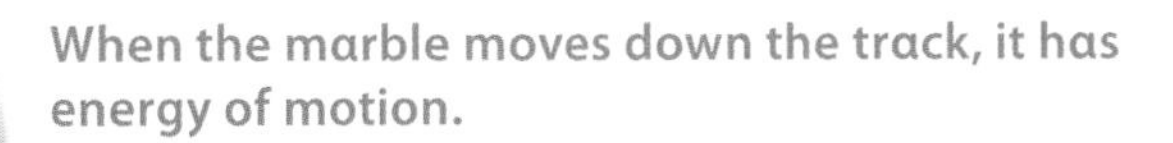
When the marble moves down the track, it has energy of motion.

What to Do, continued

3. Use a meterstick to **measure** how far the marble moves the cup. Record your observations.

4. Do steps 2 and 3 two more times.

5. Raise the upper end of the tube by taping it to the seat of a chair.

6. **Predict** how far the cup will move if you release the marble from the new height. Then do steps 2 and 3 three times from the new height.

Record

Write in your science notebook.
Use a table like this one.

my SCIENCE notebook

Energy and Height

Placement of Tube	Motion of Marble	Distance Cup Moved (cm)
Against 3 books: trial 1		

Explain and Conclude

1. Did the results support your **prediction?** Explain.
2. How did the height of the tube affect how far the marble pushed the cup?
3. **Analyze** your **data.** Use evidence to **infer** the relationship between the release height of the marble and its energy of motion.

Think of Another Question

What else would you like to find out about how an object's energy of motion changes as its position changes? How could you find an answer to this new question?

Guided Inquiry

Investigate Light

How does light move through and reflect from different materials?

Science Process Vocabulary

predict verb

When you **predict,** you tell what you think will happen.

I predict that some light will move through the leaves.

share verb

You can **share** results by telling others about your investigation and your observations.

Others can learn from my investigation if I share my results.

Materials

sheet of white paper

book

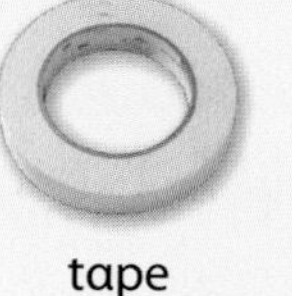

tape

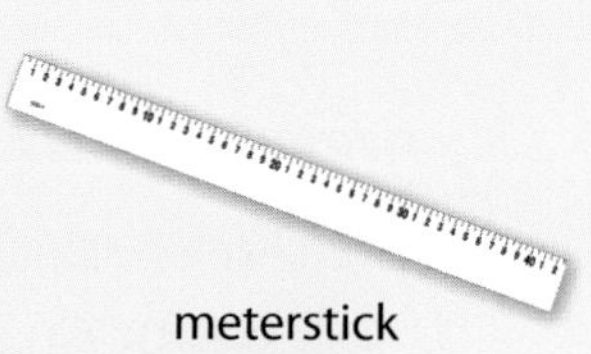

meterstick

flashlight

clear plastic wrap

group A materials

index card

marker

flat mirror

group B materials

What to Do

1. Tape a sheet of white paper to a book. Have a partner hold the book upright. Hold a flashlight about 30 cm from the book, and shine its light onto the paper. **Observe** the amount of light on the paper. Record your observations in your science notebook.

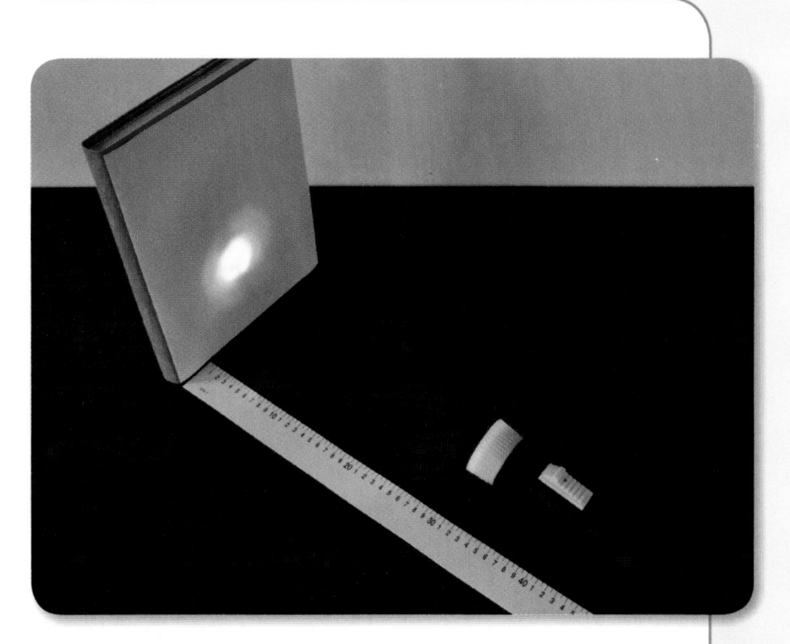

2. Tape the clear plastic wrap to the front of the flashlight. Shine the flashlight toward the paper. Record how much light moves through the plastic onto the paper.

3. Choose 3 materials from Group A to hold over the flashlight. For each material, **predict** what will happen to the amount of light on the paper if the light first moves through the material. Record your prediction. Then repeat step 2 with each material.

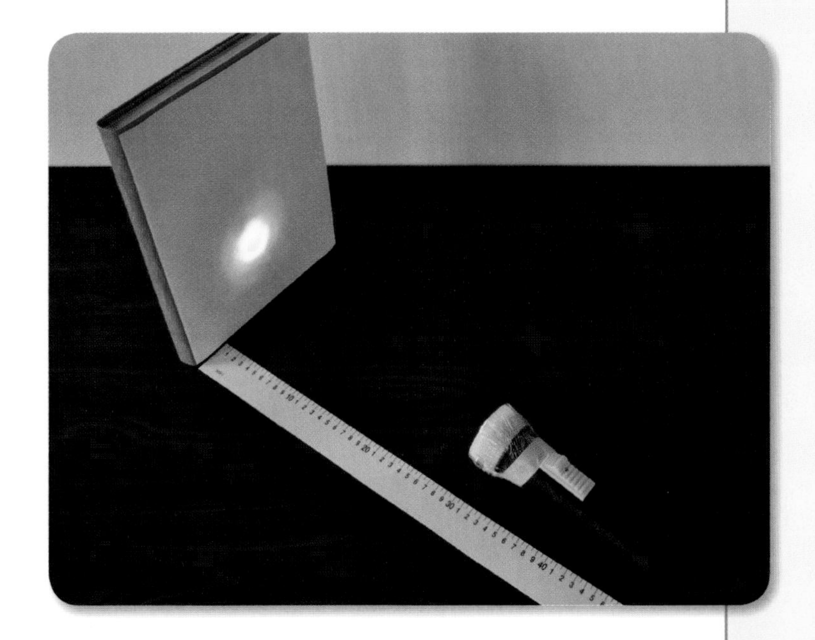

What to Do, continued

4 Write the word *BAT* on an index card.

5 Have a partner hold a flat mirror vertically on the table with the reflective side facing you.

6 Place the card in front of the mirror. The word *BAT* should be facing the mirror. Observe the reflection of the word *BAT* in the mirror. Record your observations.

7 Choose 3 materials from Group B. For each surface, first predict how the word *BAT* will appear on the surface. Then repeat step 6 with each surface.

Record

Write in your science notebook. Use tables like these.

Light Shining on Group A Materials

Material	Prediction	Observations
Flashlight alone		
Plastic wrap		

Reflection of Light by Group B Objects

Material	Prediction	Observations
Flat mirror		

Explain and Conclude

1. Do the results support your **predictions** in steps 3 and 7? Explain.
2. **Share** your results. Summarize what you learned about light in this investigation.

Think of Another Question

What else would you like to find out about how light moves through and reflects from different materials? How could you find an answer to this new question?

A funhouse mirror changes your image in many ways.

Directed Inquiry

Investigate Charged Objects

Question How does a charged balloon affect other objects?

Science Process Vocabulary

predict verb

When you **predict,** you tell what you think the results of an investigation will be.

I predict that the cloth will be attracted to the ballloon.

infer verb

When you **infer,** you use what you already know and what you observe to draw a conclusion.

I infer that rubbing caused a change in the balloon.

Materials

safety goggles

2 balloons

2 strings

tape

wool cloth

scissors

tissue paper

rubber band

aluminum foil

What to Do

1. Put on your safety goggles. Tie a piece of string to each balloon.

2. Tape the ends of both balloon strings to the edge of a table. Allow the balloons to hang freely about 3 cm apart.

The balloons do not have an electrical charge.

3. What do you think will happen to the position of the balloons if you rub each balloon with the wool cloth? Record your **prediction** in your science notebook. First rub one and then the other balloon with the cloth. Be sure to rub all parts of the balloon. Allow the balloons to hang freely. Record your **observations.**

Rubbing a balloon gives it an electrical charge.

What to Do, continued

4. Cut strips of tissue paper into small pieces. Put the pieces in a pile. Make another pile by cutting up a rubber band. Make a third pile by cutting aluminum foil into small pieces.

5. Move a balloon toward the pieces of tissue paper. Record your observations. Then rub a wool cloth against one side of the balloon. Predict what will happen if you move the balloon toward the pile of tissue paper. Then move the balloon toward the tissue paper. Record your observations.

6. Predict what will happen if you move the balloon toward the pile of rubber pieces. Record your prediction. Then repeat step 5 with the rubber pieces.

7. Predict what will happen if you move the balloon toward the pile of aluminum foil pieces. Repeat step 5 with the aluminum foil pieces.

Record

Write in your science notebook. Use a table like this one.

my SCIENCE notebook

Effect of a Charged Balloon on Objects

	Prediction	Observations
Another charged balloon		
Tissue paper pieces		

Explain and Conclude

1. **Compare** the positions of the balloons before and after you rubbed them with the cloth. What do you think caused the difference?
2. Did the tissue paper, rubber, and aluminum pieces all react the same way to the rubbed balloon? What can you **infer** about the charge on the different materials from these results?
3. Compare your group's results with the results of other groups. Explain any differences.

Think of Another Question

What else would you like to find out about how charged objects affect other objects? How could you find an answer to this new question?

A lightning strike occurs because of the effect of electrical charge.

Guided Inquiry

Investigate Electrical Conductors and Insulators

Through which materials does electrical current move?

Science Process Vocabulary

classify verb

When you **classify** objects, you sort them by their properties.

I can classify these wires by color.

investigate verb

You **investigate** when you make a plan and carry out the plan to answer a question.

I can investigate the objects to find out which are conductors and which are insulators.

Materials

safety goggles

light bulb

battery

battery holder

bulb holder

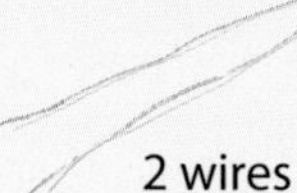

2 wires

nail

aluminum foil

index card

eraser

craft stick

paper clip

washer

cork

button

What to Do

1. Discuss with your team how you can make an electrical circuit. Your circuit must include a light bulb. Decide how to use your circuit to **investigate** whether different materials are electrical conductors or insulators.

2. Put on your safety goggles. Make your circuit. Test your circuit using the nail. The nail is a conductor, so electricity can flow in a complete loop. The light bulb should light.

3. If the bulb does not light, discuss with your team how you might improve the design of your electrical circuit. Make and test any improvements to your design.

What to Do, continued

4. **Predict** whether foil will complete the circuit and make the bulb light. Record your prediction in your science notebook. Then use the circuit to test the foil and record your **observations.**

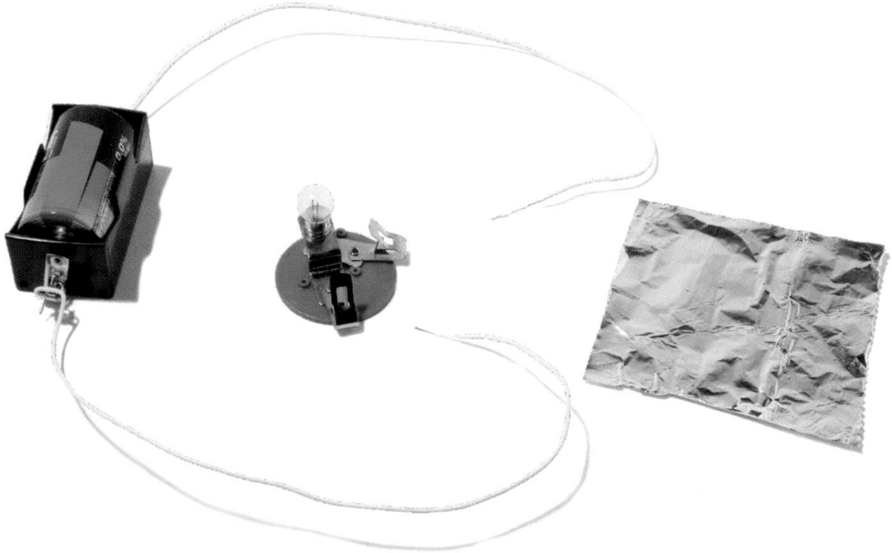

5. Choose 5 more items to test. Predict whether each item will light the bulb. Record your predictions. Then use your circuit to test the materials. Record your observations.

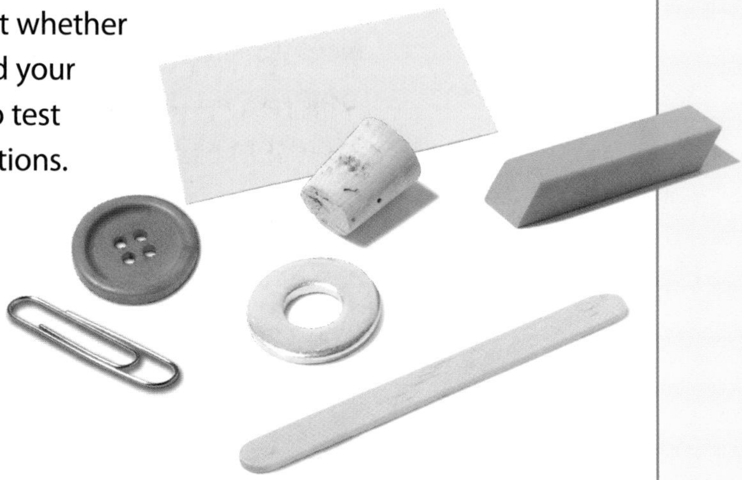

6. **Classify** each material as a conductor or an insulator. Write in your science notebook.

Record

Write in your science notebook.
Use a table like this one.

my SCIENCE notebook

Conductors and Insulators

Object	Prediction: Will the Bulb Light?	Observation: Did the Bulb Light?	Conductor or Insulator?
Aluminum foil			

Explain and Conclude

1. How do your **observations compare** with your **predictions?**
2. **Analyze** your **data.** What patterns do you see in the data?
3. **Share** your results with others. Explain any differences in the way you **classified** materials.

Think of Another Question

What else would you like to find out about electrical conductors and insulators? How could you find an answer to this new question?

Materials are chosen for safe electrical use based on whether they are insulators or conductors.

Open Inquiry

Do Your Own Investigation

Choose one of these questions, or make up one of your own to do your investigation.

- How does temperature affect how much sugar will dissolve in water?
- How does crushing an effervescent tablet affect how fast it reacts with water?
- How does the mass of a wood block affect the force you need to make it start moving?
- How does the number of pulleys affect the force needed to lift an object?
- Does heat travel faster through plastic or metal?
- Which objects can be electrically charged by rubbing them with a wool cloth?

Science Process Vocabulary

question noun

If you want to know more about something, you can ask a **question** to find out about it.

How much sugar will dissolve in the water?

Open Inquiry Checklist

Here is a checklist you can use when you investigate.

- ☐ Choose a **question** or make up one of your own.
- ☐ Gather the materials you will use.
- ☐ If needed, make a **hypothesis** or a **prediction.**
- ☐ If needed, identify, manipulate, and control **variables.**
- ☐ Make a **plan** for your **investigation.**
- ☐ Carry out your plan.
- ☐ Collect and record **data. Analyze** your data.
- ☐ Explain and **share** your results.
- ☐ Tell what you **conclude.**
- ☐ Think of another question.

The motor in a ceiling fan changes electrical energy to mechanical energy.

Write Like a Scientist

Write About an Investigation

Solubility

The following pages show how one student, Maya, wrote about an investigation. As she was stirring powdered drink mix into water, Maya was surprised that so much of the mix could dissolve in the water. She decided to do an investigation. Here is what she thought about to get started:

- Because sugar is an ingredient in the drink mix, Maya decided to test how much sugar would dissolve in water.
- She thought the solubility might depend on temperature.
- She would test 3 temperatures of water: very cold water, room-temperature water, and very warm water.
- At each temperature, she would stir sugar in the water until the sugar dissolved. She would continue adding small amounts at a time until no more sugar would dissolve.

Dissolving Sugar

Undissolved sugar in water

Dissolved sugar in water

Model

Question

How does temperature affect how much sugar will dissolve in water?

Be sure your question can be answered by an investigation.

Materials

graduated cylinder (100 mL)

water: very warm, room-temperature, and very cold

3 plastic cups (10 oz)

tape

thermometer

500 g sugar

plastic spoon

List all the materials you will need.

my SCIENCE notebook Your Investigation

Now it's your turn to do your investigation and write about it. Write about the following checklist items in your science notebook.

- [ ] **Choose a question or make up one of your own.**
- [ ] **Gather the materials you will use.**

Model

My Hypothesis

If I decrease the temperature of water, then less sugar will dissolve in the cold water. If I increase the temperature of water, then more sugar will dissolve in the very warm water.

You can use "If..., then..." statements to make your hypothesis clear.

Cold

my SCIENCE notebook

Your Investigation

☐ **If needed, make a hypothesis or prediction.**

Write your hypothesis or prediction in your science notebook.

Model

Variable I Will Change

I will use water with 3 different temperatures: very warm, room-temperature, and cold.

Variable I Will Observe or Measure

I will observe how much sugar dissolves in the water.

Variables I Will Keep the Same

Everything else will be the same. Each test will have the same amount of water. Each will have the same type of sugar.

Answer these three questions:

1. What one thing will I change?
2. What will I observe or measure?
3. What things will I keep the same?

my SCIENCE notebook

Your Investigation

☐ **If needed, identify, manipulate, and control variables.**

Write about the variables for your investigation.

Model

My Plan

1. Label the cups with tape: **Warm**, **Room-Temperature**, and **Cold**.
2. Pour 50 mL of room-temperature water into the **Room-Temperature** cup. Measure and record the temperature.
3. Mix 1 spoonful of sugar into the water. Stir well. Continue adding sugar, 1 spoonful at a time, until the sugar no longer dissolves. Record the number of spoonfuls that dissolve in the room-temperature water. Rinse and dry all supplies.
4. Repeat steps 2 and 3 using very cold water.
5. Repeat steps 2 and 3 again using very warm water.

Reread your plan to make sure you did not leave out any important steps.

Your Investigation

☐ **Make a plan for your investigation.**

Write the steps for your plan.

Model

my SCIENCE notebook Your Investigation

☐ **Carry out your plan.**

Be sure to follow your plan carefully.

Model

Data (My Observations)

Observations of Sugar in Water

Cup	Water Temperature (°C)	Sugar Added (Spoonfuls)
Warm	83°	14
Room-Temperature	19°	8
Cold	5°	6

Use a table to record your data.

My Analysis

The least amount of sugar dissolved in cold water.
The greatest amount of sugar dissolved in very warm water.

Explain what happened based on the data you collected.

my SCIENCE notebook

Your Investigation

☐ **Collect and record data. Analyze your data.**

Collect and record your data, and then write your analysis.

Model

Scientists often share results so others can find out what was learned.

How I Shared My Results

I made a presentation to the class. First I stirred some sugar in water to demonstrate dissolving. Next I shared the results from the investigation. Then I told how my investigation related to stirring powdered drink mix into water.

Tell what you conclude and what evidence you have for your conclusion.

My Conclusion

Less sugar dissolves in water if the water is cooler than room-temperature water. More sugar dissolves if the water is warmer than room-temperature water.

Investigations often lead to new questions for Inquiry.

Another Question

I wonder if the amounts of other substances, such as salt, that dissolve in water at different temperatures would be the same as they are for sugar.

my SCIENCE notebook

Your Investigation

- ☐ Explain and share your results.
- ☐ Tell what you conclude.
- ☐ Think of another question.

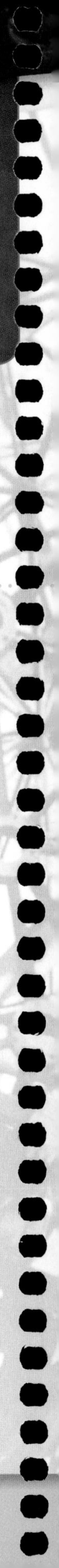

Science and Technology

The Keck telescopes allow astronomers to see two colliding galaxies nearly 5 billion light-years away.

How Technology Helps Scientists

Technology helps scientists to discover new information and to make people's lives better. Modern telescopes, digital computers, and electronic microscopes allow scientists to make better observations and measurements than in the past.

Telescopes An optical telescope is a system of lenses or mirrors that collects light from distant objects. Telescopes allow observers to see fainter, more distant objects than they can see with only their eyes. Scientists today use telescopes to investigate the age of the universe, observe the life cycles of stars, and look for planets outside our solar system. Telescopes help scientists learn more about space.

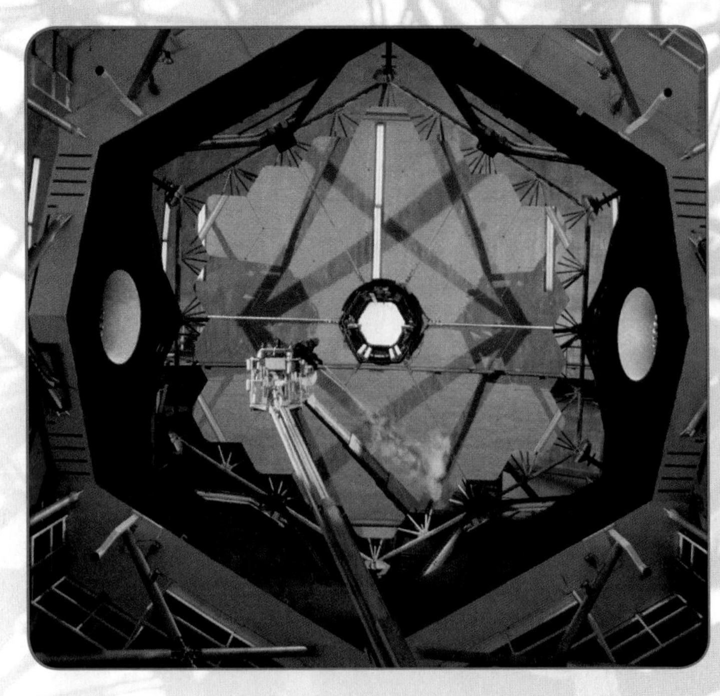

Each of the twin Keck telescopes has a 10-meter system of mirrors. They are the world's largest optical telescopes.

Digital Computers Scientists use digital computers to collect and store data, make calculations, and create models. Since the middle of the twentieth century, digital computers have been changing our world. They contribute to saving lives with medical equipment, navigating jet planes, and forecasting weather. Computers allow us to use email, the Internet, and television. Banks, stores, and hospitals depend on computers that store and share data. Video games and movies are produced with the help of computers. Tiny computers are in appliances, watches, phones, and toys. Digital computers affect many things in our work and play.

This ENIAC (Electronic Numerical Integrator and Computer) from the 1940s was the first electronic digital computer in the United States. It filled a 9-by-18-meter room. A laptop computer today is more powerful than ENIAC was.

Microscopes Many microscopes rely on lenses to make small objects appear larger. However, the atom probe field ion microscope, or atom probe, works without lenses. It makes the tiniest particles of matter visible. The atom probe allows scientists to examine a material one atom at a time! They can use this technology to make new materials with improved properties. For example, scientists can experiment with different elements to produce a new material that is strong and resistant to rust. This material can be used to make jet turbine blades.

A compound is examined with an atom probe field ion microscope. Each dot in the photo represents a single atom!

SUMMARIZE

What Did You Find Out?

1. How do scientists use telescopes?
2. What are three ways society might be different if we did not have digital computers?
3. How does the ability to examine a single atom help scientists?

Identify Uses of Computer Technology

Look at the list of items that use technology. Choose two items. Explain how the items are used. Then write a few paragraphs to explain how your life would be different if you did not have these items.

Computer Technology

Examples	How Item Is Used
Cell phone	
Television	
Digital camera	
Car GPS system	

These are just a few everyday items that are controlled by computers.

ACKNOWLEDGMENTS Grateful acknowledgment is given to the authors, artists, photographers, museums, publishers, and agents for permission to reprint copyrighted material. Every effort has been made to secure the appropriate permission. If any omissions have been made or if corrections are required, please contact the Publisher.

ILLUSTRATOR CREDITS **20** Paul Dolan. **196** Paul Dolan. All maps by Mapping Specialists.

PHOTOGRAPHIC CREDITS **Front, Back Cover** Alaska Stock Images/National Geographic Image Collection. **1-2, 3-4, 5-6, 7-8** Alaska Stock Images/National Geographic Image Collection. **10** (c) Kletr/Shutterstock. (b) marinini/Shutterstock. **11** (tl) Johan Swanepoel/Shutterstock. (tc) Wildlife/Peter Arnold, Inc.. (tr) Dejan Ljamić/iStockphoto. (bc) Michael Durham/Minden Pictures/National Geographic Image Collection. (br) Arco Images/Reinhard, H./Alamy Images. **13** (t) Millard H. Sharp/Photo Researchers, Inc.. **14** (c) JLP/Jose L. Pelaez/Ivy/Corbis. (b) Mikael Kjellstrom/Alamy Images. **14-15** (t) Creatas/Jupiterimages. **15** (bl) FLPA/David Hosking/age fotostock. (br) 10" x 14" (35 MB - 300ppi RGB)/age fotostock. **16** (t) Creatas/Jupiterimages. (b) Doug Perrine/Nature Picture Library. **17** (b) stephan kerkhofs/Shutterstock. **18** (c) Hill Street Studios/Blend Images/Corbis. (bl) Karl E. Deckart/Phototake. (br) John Durham/Photo Researchers, Inc.. **18-19** (t) Wim van Egmond/Visuals Unlimited. **20** (t) Wim van Egmond/Visuals Unlimited. (c) Runk/Schoenberger/Grant Heilman Photography. (b) Barry Runk/Stan/Grant Heilman Photography. **21** (b) Biophoto Associates/Photo Researchers, Inc.. **22** (t) JuliaSha/Shutterstock. (b) Dave King/Dorling Kindersley/Getty Images. **24** (t) JuliaSha/Shutterstock. **25** (b) Raymond Gehman/National Geographic Image Collection. **26** (t) John Foxx/White/Photolibrary. (b) Jonathan Noden-Wilkinson/Shutterstock. **27** (b) John Foxx/White/Photolibrary. **28** (t) John Foxx/White/Photolibrary. **29** (b) James P. Blair/National Geographic Image Collection. **30** (t) Wayne Stadler/iStockphoto. **32** (t) Wayne Stadler/iStockphoto. **33** (b) mihalec/Shutterstock. **34** (t) Creatas/Jupiterimages. (cl) Vasko Miokovic/iStockphoto. (cr) Ana Amorim/iStockphoto. (b) Hermínia Lúcia Lopes Serra/Shutterstock. **35** (t) Creatas/Jupiterimages. **36** (t) Creatas/Jupiterimages. **37** (b) Julius Elias/Shutterstock. **38** (t) White/Jason Reed/Photolibrary. (b) Suzanne Carter-Jackson/iStockphoto. **41** (t) White/Jason Reed/Photolibrary. (b) Suzanne Carter-Jackson/iStockphoto. **45** (b) Mark Newman/SuperStock. **46** (c) blickwinkel/Alamy Images. **49** (b) Color-Pic/Animals Animals. **50** (t) Joel Sartore/National Geographic Image Collection. (b) Shiyana Thenabadu/Alamy Images. **52** (t) Joel Sartore/National Geographic Image Collection. **53** (b) Digital Vision/Getty Images. **54** (t) Colin Underhill/Alamy Images. (c) Fancy/Alamy Images. **56** (t) Colin Underhill/Alamy Images. **57** (b) BananaStock/Jupiterimages. **58** (c) sajko/Shutterstock. **58-59** (t) Robert J. Erwin/Photo Researchers, Inc.. **59** (b) Barbara Henry/iStockphoto. **60** (c) Jerry And Marcy Monkman / Danita Delimont/Alamy Images. **60-67** (t) Alan & Linda Detrick/Photo Researchers, Inc.. **68** (t) Emelyanov/Shutterstock. (b) Blend Images/Alamy Images. **69** vario images GmbH & Co.KG/Alamy Images. **70** (l) image100 Science A/Alamy Images. (r) Bill Aron/PhotoEdit. **71** (bg) Pascal Goetgheluck/Photo Researchers, Inc.. (t) Emelyanov/Shutterstock. (b) Michael Newman/PhotoEdit. **72-73** (t) DigitalStock/Corbis. **73** (bg) Creatas/Jupiterimages. **76** (t) Photodisc Collection/White/Photolibrary. (c) EJ Carr/81a/Photolibrary. **78** (tl) Photodisc Collection/White/Photolibrary. **79** (b) Felix Stensson/Alamy Images. **80** (t) PhotoDisc/Getty Images. **80-81** (t) John Foxx Images/Imagestate. **82** (t) PhotoDisc/Getty Images. **84** (c) Brand X Pictures/Jupiterimages. **84-85, 86** (t) R. Gilmozzi, Space Telescope Science Institute/European Space Agency; Shawn Ewald, JPL; and NASA. **87** (b) NASA Human Space Flight Gallery. **88** (t) NASA, ESA, and the Hubble Heritage (STScI/AURA)-ESA/Hubble Collaboration. (b) Mark Garlick/Photo Researchers, Inc.. **89** (bg) NASA, ESA, and the Hubble Heritage (STScI/AURA)-ESA/Hubble Collaboration. (t) PhotoDisc/Getty Images. (b) Stockbyte/Getty Images. **90** (bg) PhotoDisc/Getty Images. (l) NASA/JPL. (r) Science Source/Photo Researchers, Inc.. **91** (t) NASA, ESA, and the Hubble Heritage (STScI/AURA)-ESA/Hubble Collaboration. (b) The International Astronomical Union/Martin Kornmesser. **92** (b) Corbis Premium RF/Alamy Images. **95** (b) PhotoDisc/Getty Images. **96** (b) Wildlife/Peter Arnold, Inc.. **96-97** (t) Janaka Dharmasena/Shutterstock. **99** (b) sodapix sodapix/F1Online RF/Photolibrary. **104** (b) Vanessa Nel/Shutterstock. **104-105** (t) Creatas/Jupiterimages. **106** (t) Creatas/Jupiterimages **107** (b) ultimathule/Shutterstock. **108** (c) IMAGEMORE Co., Ltd./Alamy Images. **108-109** (t) Françoise Emily/Alamy Images. **112-113** (t) Bruce Dale/National Geographic Image Collection. **114** (t) Bruce Dale/National Geographic Image Collection. **115** (b) joyfull/Shutterstock. **116** (t) HomeStudio/Shutterstock. (b) Jan Tadeusz/Alamy Images. **118** (t) HomeStudio/Shutterstock. (b) Image Source/Jupiterimages. **119** (b) NOAA/Getty Images. **120-121** (t) Andy Cash/Shutterstock. **121** (b) tanOd/Shutterstock. **124-127** (t) DigitalStock/Corbis. **128** (t) Dick Stada/Shutterstock. **128-129** (t) DigitalStock/Corbis. **130** (b) Brian J. Skerry/National Geographic Image Collection. **131** (b) Pallava Bagla/Corbis. **132-133** (t) John Foxx Images/Imagestate. **132** Javier Larrea/age fotostock. **138** (cl) Peter Arnold, Inc./Alamy Images. (b) Beata Jancsik/Shutterstock. **138-139, 140** (t) Sebastian Duda/Shutterstock. **141** (b) Martin Allinger/Shutterstock. **142** (b) Karen Struthers/Shutterstock. **142-143, 144** (t) PhotoDisc/Getty Images. **145** (b) BrandX/Jupiterimages. **146-147, 148** (t) PhotoDisc/Getty Images. **149** (b) Polka Dot Images/Jupiterimages. **150-151, 152** (t) John Foxx Images/Imagestate. **153** (b) PhotoDisc/Getty Images. **154** (t) Image Source Black/Alamy Images. (b) Sam Diephuis/zefa/Corbis. **157** (t) Image Source Black/Alamy Images. **161** (b) Alistair Berg/Jupiterimages. **162,164** (t) delihayat/Shutterstock. **165** (b) PhotoDisc/Getty Images. **169** (b) Benoit Rousseau/iStockphoto. **170, 172** (t) freelion/Shutterstock. **173** (b) W. Robert Moore/National Geographic Image Collection. **174** (c) Steve Walsh/Alamy Images. (b) Rovenko Design/Shutterstock. **174-175** (t) Francois Etienne du Plessis/Shutterstock. **175** (b) DenisKlimov/Shutterstock. (c) Eduardo Rivero/Shutterstock. **176** (b) Vladislav Ociacia/iStockphoto. **176-177** (t) Francois Etienne du Plessis/Shutterstock. **178-179** (t) Andy Piatt/Shutterstock. **180** (t) Andy Piatt/Shutterstock. **181** (b) John Foxx Images/Imagestate. **182-183, 184** (t) Stephen Aaron Rees/Shutterstock. **185** (b) INSADCO Photography/Alamy Images. **186-187, 188** (t) Svetlana Yudina/Shutterstock. **189** (b) PhotoDisc/Getty Images. **190** (cl) PhotoDisc/Getty Images. (bl) Artville. (bcl) ra-design/Shutterstock. (bcr) Ariusz Nawrocki/Shutterstock. (br) Mudassar Ahmed Dar/Shutterstock. **190-191** (t) Dick Stada/Shutterstock. **193** (b) John Foxx Images/Imagestate. **194** (b) Steve Holderfield/Shutterstock. **194-195** (t) John Foxx Images/Imagestate. **195** (b) XPhantom/Shutterstock. **196-203** (t) Kathie Nichols/Shutterstock. **204** (c) Hubble Space Telescope/W.M. Keck Observatory. (b) Roger Ressmeyer/Corbis. **204-05** (t) Alexey Arkhipov/Shutterstock. **204-205** (bg) photodisc/age fotostock. **205** (b) Corbis. **206** (b) Dr. G.K.L. Cranstoun/Photo Researchers, Inc.. **206-207** (t) Alexey Arkhipov/Shutterstock. **207** (cl) Doroshin Oleg/Shutterstock. (cr) Seth Joel/Getty Images. (cr inset) Anobis/Shutterstock. (bl) Olaru Radian-Alexandru/Shutterstock. (br) Jaroslaw Grudzinski/Shutterstock.

Neither the Publisher nor the authors shall be liable for any damage that may be caused or sustained or result from conducting any of the activities in this publication without specifically following instructions, undertaking the activities without proper supervision, or failing to comply with the cautions contained herein.

PROGRAM AUTHORS Judith Sweeney Lederman, Ph.D., Director of Teacher Education and Associate Professor of Science Education, Department of Mathematics and Science Education, Illinois Institute of Technology, Chicago, Illinois; Randy Bell, Ph.D., Associate Professor of Science Education, University of Virginia, Charlottesville, Virginia; Malcolm B. Butler, Ph.D., Associate Professor of Science Education, University of South Florida, St. Petersburg, Florida; Kathy Cabe Trundle, Ph.D., Associate Professor of Early Childhood Science Education, The Ohio State University, Columbus, Ohio; David W. Moore, Ph.D., Professor of Education, College of Teacher Education and Leadership, Arizona State University, Tempe, Arizona

THE NATIONAL GEOGRAPHIC SOCIETY
John M. Fahey, Jr., President & Chief Executive Officer
Gilbert M. Grosvenor, Chairman of the Board

Copyright © 2011 The Hampton-Brown Company, Inc., a wholly owned subsidiary of the National Geographic Society, publishing under the imprints National Geographic School Publishing and Hampton-Brown.

All rights reserved. No part of this book may be reproduced or transmitted in any form or by any means, electronic or mechanical, including photocopying, recording, or by an information storage and retrieval system, without permission in writing from the Publisher.

National Geographic and the Yellow Border are registered trademarks of the National Geographic Society.

National Geographic School Publishing
Hampton-Brown
www.NGSP.com

Printed in the USA. RR Donnelley, Menasha, WI

ISBN: 978-0-7362-7817-1

12 13 14 15 16 17 18 19 20

7 8 9 10